Green Chemistry

Greener Alternatives to Synthetic Organic Transformations

Green Chemistry

Greener Alternatives to Synthetic Organic Transformations

V.K. Ahluwalia

Alpha Science International Ltd.
Oxford, U.K.

Green Chemistry
264 pgs.

V.K. Ahluwalia
Honorary Visiting Professor
Dr. B.R. Ambedkar Centre for Biomedical Research
Former Professor of Chemistry
University of Delhi
Delhi

ALPHA SCIENCE INTERNATIONAL LTD.
7200 The Quorum, Oxford Business Park North
Garsington Road, Oxford OX4 2JZ, U.K.

www.alphasci.com

ISBN 978-1-84265-650-1

Printed in India

Preface

Green chemistry deals with environmentally benign organic transformations. The book is divided into two parts.

Part I deals with introduction, principles of green chemistry and planning a green synthesis

Part II deals with synthetic organic transformations in aqueous phase, solid phase, photochemical transformation, transformations using phase transfer catalysts, using sonication, microwaves, enzymatic transformations, in ionic liquids, polyethylene glycol and also transformations using polymer supported reagents or substrates. Besides the above, some miscellaneous transformations are also described. The experimental details of a large number of transformations in all the above heads are given.

This book will be extremely helpful to students at undergraduate and postgraduate levels and also to research students, chemists and scientists engaged in research in the Colleges, Universities and Industries.

V.K. Ahluwalia

Preface

Green chemistry deals with environmentally benign organic transformations. The book is divided into two parts.

Part I deals with introduction, principles of green chemistry and planning a green synthesis.

Part II deals with various green transformations in aqueous phase, solid phase, photochemical transformation, transformations using phase transfer catalysts, using sonication, microwaves and ionic liquids, polyethylene glycol and its derivatives, using polymer supported reagents/substrates. Besides the above, some miscellaneous transformations are also described. The experimental details of a large number of transformations of the above reactions are given.

This book will be extremely helpful to students at undergraduate and postgraduate levels and also to research students, chemists and scientists engaged in research in the Colleges, Universities and Industries.

V.K. Ahluwalia

Contents

PART I

INTRODUCTION

In this part following are described

– Principals of green chemistry
– How to plan a green synthesis

Chapter 1

Introduction

Chemistry dealing with environmentally benign chemical synthesis is termed green chemistry. It deals with chemical processes that reduce the use and generation of hazardous substances or by products; all these products are basically responsible for pollution of the environment. Due to this, it is important to bring about changes in the chemistry curriculum in the universities, colleges and schools. In fact, the aim should be to educate the students at various levels about the need of green chemistry.

1.1 PRINCIPLES OF GREEN CHEMISTRY

As already stated, green chemistry deals with chemical synthesis which are benign. This is one of the important ways to prevent or reduce the pollution. Following are given the twelve principles of green chemistry*.

1. It is better to prevent waste than to treat or clean up waste after it is formed.
2. Synthetic methods should be designed to maximize the incorporation of all materials used in the process into the final product.
3. Wherever practicable, synthetic methodologies should be designed to use and generate substances that possess little or no toxicity to human health and the environment.
4. Chemical products should be designed to preserve efficacy of function while reducing toxicity.
5. The use of auxiliary substances (solvents, separation agents, etc.) should be made unnecessary whenever possible and, when used, innocuous.
6. Energy requirements should be recognized for their environmental and economic impacts and should be minimized. Synthetic methods should be conducted at ambient temperature and pressure.
7. A raw material or feedstock should be renewable rather than depleting whenever technically and economically practical.

* Paul T. Anastas and John C. Warner, Green Chemistry, Theory and Practice, Oxford Unicersity Press, New York, 1988.

8. Unnecessary derivatization (blocking group, protection/deprotection, temporary modification of physical/chemical processes) should be avoided whenever possible.
9. Catalytic reagents (as selective as possible) are superior to stoichiometric reagents.
10. Chemical products should be designed so that at the end of their function they do not persist in the environment and instead break down into innocuous degradation products.
11. Analytical methodologies need to be further developed to allow for real-time in-process monitoring and control prior to the formation of hazardous substances.
12. Substances and the form of a substance used in a chemical process should be chosen so as to minimize the potential for chemical accidents, including releases, explosions, and fires.

1.2 HOW TO PLAN A GREEN SYNTHESIS

While planning a green synthesis, following points should be considered.

1. The synthetic methods should be such that all the starting materials be converted into the final product.

It is normally believed that if more of a starting material gives one mole to the product, the yield is 100%. Hourever, such a synthesis may generate of byproducts or waste, which is not visible. Though such a synthesis is 100% (as per the yield mentioned above), it is not considered to be a green synthesis. For example, Hofmann elimination reaction, Witting reaction and Grignard reaction may proceed with 100% yield but they do not take into account the large amounts of byproducts produced. Some examples are given below.

$$CH_3CHO \xrightarrow[2)\ H_3O^+]{1)\ CH_3MgI} \underset{\text{Product}}{CH_3-\overset{\overset{\displaystyle H}{|}}{\underset{\underset{\displaystyle CH_3}{|}}{C}}-OH} + \underset{\text{Byproduct}}{MgI(OH)} \quad \text{(Grigrand reaction)}$$

$$R_2C=O + Ph_3P=CHR' \longrightarrow \left[\begin{array}{ccc} Ph_3\text{—}P^+ & & O^- \\ | & & | \\ R'\text{—}C & \text{—} & C\text{—}R \\ | & & | \\ H & & R \end{array}\right] \quad \text{(Wittig reaction)}$$

$$\downarrow$$

$$\underset{\text{Product}}{R'-CH=CR_2} + \underset{\text{Byproduct}}{O=PPh_3} \longleftarrow \left[\begin{array}{ccc} Ph_3P^+ & \text{—} & O \\ | & & | \\ R'\text{—}C & \text{—} & C\text{—}R \\ | & & | \\ H & & R \end{array}\right]$$

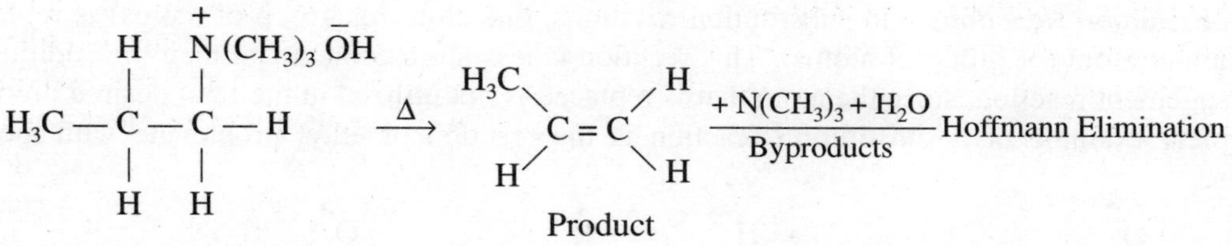

$$H_3C-\underset{H}{\overset{H}{C}}-\underset{H}{\overset{\overset{+}{N}(CH_3)_3\,\overline{O}H}{C}}-H \xrightarrow{\Delta} \underset{\text{Product}}{(H_3C)(H)C=C(H)(H)} + \underset{\text{Byproducts}}{N(CH_3)_3 + H_2O} \quad \text{Hoffmann Elimination}$$

The extent of incorporation of the starting material into the product is by the following formula [R.A. Sheldon Chem. Ind. (London), 1992, 903]

$$\%\ \text{Atom economy} = \frac{\text{F.W of atoms utilised}}{\text{F.W of the reactants used in the reaction}} \times 100$$

The calculation of the atom economy will be well understood by considering some of the common reactions like addition reactions, rearrangement reactions, substitution reactions and elimination reactions, which we generally come across in organic synthesis

(a) Addition Reactions Consider the bromination of 1-butene.

$$\underset{\text{1-Butene}}{CH_3\,CH_2\,CH=CH_2} + Br_2 \xrightarrow{CCl_4} \underset{\text{1, 2-Dibromobutane}}{CH_3\,CH_2\,CH\,Br\,CH_2\,Br}$$

In the above reaction, all the atoms of starting materials (1-butene and bromine) are incorporated into the final product (1, 2-dibromobutane). So the reaction is 100% atom economical reaction.

In a similar way cycloaddition reaction (Diels-Alder reaction) of butadiene and ethyne and addition of hydrogen to olefins are 100% atom economical reactions

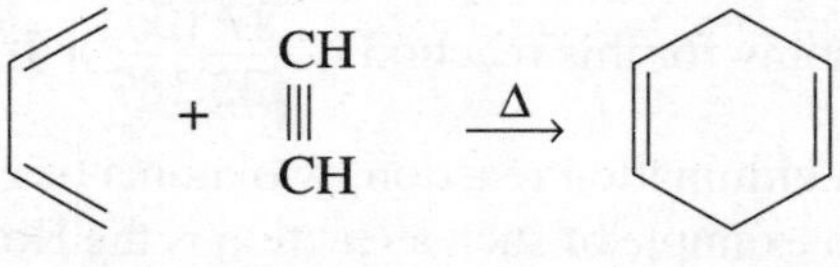

$$\text{Butadiene} + \underset{\text{Ethyne}}{CH\equiv CH} \xrightarrow{\Delta} \text{1, 4-cylo hexadiene}$$

$$\underset{\text{1-Butene}}{CH_3\,CH_2\,CH=CH_2 + H_2} \xrightarrow{Ni} \underset{\text{Butane}}{CH_3\,CH_2\,CH_2\,CH_3}$$

(b) Rearrangement Reactions In rearrangement reactions, there is rearrangement of atoms.

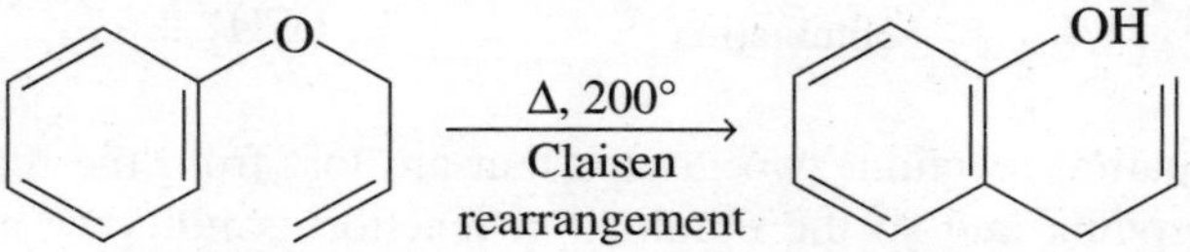

$$\text{Allyl phenylether} \xrightarrow[\text{rearrangement}]{\Delta,\ 200^\circ\ \text{Claisen}} \text{O-Allylphenol}$$

The rearrangement reaction like addition reactions is also 100% atom economical, since all the atoms in the reactant are incorporated into the product.

(c) Substitution Reactions In substitution reactions, one atom (or group of atoms) is replaced by another atom (or group of atoms). This reaction is less atom-economical than the addition or rearrangement reaction, since the atom that is replaced is not utilized in the final desired product. A typical example of a substitution reaction is the reaction of ethyl propionate with methyl amine.

$$\underset{\text{Ethyl propionate}}{CH_3CH_2\overset{\overset{\Large O}{\|}}{C}-OC_2H_5} + \underset{\text{Methyl amine}}{H_3C-\overset{\overset{\Large H}{|}}{N}-H} \longrightarrow \underset{\text{N-Methyl propamide}}{H_3C\,CH_2-\overset{\overset{\Large O}{\|}}{C}-NHCH_3} + \underset{\text{Ethyl alcohol}}{CH_3CH_2OH}$$

As seen, in the above substitution reaction besides one hydrogen atom on the amine, the leaving group ($OCH_2\,CH_3$) is also not utilized is the desired N-methyl propamide. The remaining atoms of the reactants are incorporated into the final product. The total of atomic weights of the atoms of the reactants that are incorporated is 87.120 g/mol, and the total molecular weight of the reagent used is 133.189 g/mol. Thus a molecular weight of 46.069 g/mol (133.189 – 87.120) are not utilised in the reaction (see table below)

	Reagent		*Utilized*		*Unutilized*	
	Formula	*FW*	*Formula*	*FW (g/mol)*	*Formula*	*FW (g/mol)*
	$C_5H_{10}O_2$	102.132	C_3H_5O	57.057	C_2H_5O	45.061
	CH_5N	31.057	CH_4N	30.049	H	1.008
Total	$C_6H_{15}NO_2$	133.189	C_4H_9NO	87.106	C_2H_6O	46.069

Therefore the percentage economy for this reaction is $\dfrac{87.106}{133.189} \times 100 = 65.41\%$

(d) Elimination Reactions In an elimination reaction, two atoms or group of atoms are lost from the substrate to form a π bond. An example of such a reaction is the Hoffmann elimination reaction as given below:

$$H_3C-CH(H)-CH_2-\overset{+}{N}(CH_3)(H_3C)-CH_3 \xrightarrow[\text{Hoffmann elimination}]{\bar{O}H,\ \Delta} H_3C-CH{=}CH_2 + (H_3C)(H_3C)N-CH_3 + H_2O$$

In the above elimination reaction, two groups that are lost from the reaction are not present in the final desired product and so the elimination reaction is not very atom-economical. The percentage atom economy is 35.30%

2. Unnecessary derivation (blocking group protection/deprotection) should be avoided whenever possible.

The use of protecting group (blocking group) is also a factor in the atom economy of a reaction. Though sometimes it is necessary to use protecting groups particularly to solve a chemoselectivity problem, these should be added to the reaction mixture in stoichiometric amount and then removed after completion of the reaction

As an example the reaction of methyl cyclohexanone-2 carboxylate with grignard reagent in order to get the required product, it is necessary to use 1, 2-ethane diol to protect the keto group from reacting with the grignard reagent.

Methyl cyclohexanone 2-carboxylate + HO–CH₂CH₂–OH (1, 2-Ethane diol) →

$\xrightarrow{2\ CH_3MgBr}$ (O⁻ Mg⁺ Br, CH_3, CH_3) $\xrightarrow{H_3O^+}$ Product (OH, CH_3, CH_3) + HO–CH₂CH₂–OH (1,2-Ethane diol)

Another example of protection is the conversion of m-aminobenzonitrile into m-aminobenzaldehyde. In this case, the amino group has to be protected by benzylation. This is followed by stevens reaction and final deprotection.

m-Aminobenzonitrile (NH_2, C≡N) $\xrightarrow{C_6H_5CH_2Cl}$ ($NHCH_2C_6H_5$, C≡N) $\xrightarrow[SnCl_2]{HCl}$ ($NHCH_2C_6H_5$, CHO) ⟶

$\xrightarrow{deprotection}$ m-Aminobenzaldehyde (NH_2, CHO)

3. Use an environmentally benign solvent.

A number of solvents like methylene chloride, chloroform, perchloroethylene (PCE), carbon tetrachloride and aromatic hydrocarbons (like benzene, toluene etc) are generally used in reactions due to their excellent solvent properties. It is, however known that the halogenated solvents (like CH_2Cl_2, $CHCl_3$, PCE, CCl_4 etc.) have been identified as suspected human carcinogens. Besides, these solvents are responsible for environmental pollution particularly destroying the ozone layer leading to the formation of ozone hole. The solvents which are responsible for destorying the ozone layer also include, chlorofluorocarbons (CFCs)

In view of the above, it is necessary to carry out the reaction in environmentally benign solvents like water, super critical carbon dioxide, ionic liquids and polyethylene glycol.

The use of water as a solvent for organic reaction has been known to people till about the mid of twentieth century. Water is the cheapest solvent available and its use does not cause problems of pollution, which is a major concern in using volatile organic solvents. According to C & EN news of Sept. 3, 2007, it is recommended that in case organic solvents fail, try water.

Water has been used as such as a solvent in a number of reactions. It has also been used as super critical water (or in Near Water (NCW) Region). The critical temperature of super critical water is 374°C and 22.1 MPa. Due to its unique physical and chemical properties that are quite different than those of ambinet water, it has been used as a medium for organic reactions. The near original water region is described as 250-300° at pressure 100-80 bar [For more details about organic reactions in super critical or near water (NCW) region see, V.K. Ahluwalia and R.S. Varma, Green solvents for organic synthesis, Narosa Publishing house, 2009, Chapter 2 and the references cited there in]. Besides this, microwave assisted organic reactions and biocatalytic reactions have also been performed in water.

Super critical carbon dioxide has also been used as a solvent in organic reaction. Carbon dioxide, as we know can exist in different states depending on the temperature and pressure of its surrounding. If we increase the pressure, CO_2 becomes liquid; at this point (–56°C and 5.1 atm), CO_2 exist simultaneously as gas, liquid and solid. However, at 31°C and 73 atm. CO_2 exists as super critical fluid. In super critical state, CO_2 has a viscosity similar to that of a gas and density similar to that of a liquid. The solvent properties of SC-CO_2 (e.g. dielectric constant, solubility parameters, viscosity and density) can be altered or charged in a manner not possible with conventional solvents via manupulation of temperature and pressure. The properties of SC-CO_2 are intermediate between that of liquid and gas. [For more details about super critical CO_2, see Green solvents for organic synthesis V.K. Ahluwalia and R.S. Varma, Narosa Publishing House, 2009, Chapter 5 and the references cited there in].

Ionic liquids are emerging as novel replacement for volatile organic compounds used traditionally as industrial solvents and reduce the volatility, environmental and human health are safety concerns that are experienced by exposure to organic solvents. Ionic liquids are made up of at least two components which can be varied (the cation and anion). These are liquid at ambient temperature and are colourless, have low viscosity and can be easily handled. In fact, these have attractive properties for a solvent. The solvent can be designed with a particular end use in mind or to possess particular set of properties. Hence these are also known as 'designer solvents'. Properties such as melting point, viscosity density and hydrophobicity can be varied by simple charges to the structure of ions. These can be recycled and this leads to a reduction of the cost of the process.

The reactions are often quicker and easy to carry out than in conventional organic solvents. These are used to enhance activity, selectivity and stability of transition metal catalyst. The ionic liquids have virtually no vapor pressure and posses good thermal stability [For more details about ionic liquids see V.K. Ahluwalia and R.S. Varma, Green solvents for organic synthesis, Narosa Publishing House, 2009, Chapter 7 and the references cited there in].

Polyethylene glycol (PEG) and its solutions are believed to be a green reaction medium of the future. Polyethylene glycol, PEG, HO-$(CH_2\ CH_2O)_n$-H, is available in a variety of molecular weights. A special characteristics of PEG is that it has low flammability and is biodegrable. It is stable to acid, base and high temperatures and can be recovered and recycled. It has been used as a solvent for a number of organic reactions [For details about PEG, see V.K. Ahluwalia and R.S. Varma, Green solvents for organic synthesis, Narosa Publishing house, 2009, Chapter 8 and the references cited there in].

4. The requirement of energy should be minimised to a bare minimum.

It is well known that energy is required for any chemical reaction. This requirement should be kept to a bare minimum. For example, if the synthesis (the starting materials and reagents) are soluble in a particular solvent, the reaction mixture has to be heated to reflux for completion of a reaction, it should be kept in mind, that the time required for completion of the reaction should be minimum, so that minimum amount of energy is required. It is, however known that use of a catalyst is beneficial for lowering the requirement of energy for a reaction.

In cast, the final product has to be purified by crystallisation, distillation etc, energy is required. The chemical process be so designed that there is no need for purification or separation.

It is now well known the energy requirement can be kept bare minimum by using microwaves, sonication or photoactivation.

Among the important tools for organic synthesis, microwaves as alternative source of energy is being used for organic reactions. The household and industrial microwave oven, operate at a fixed frequency at 2.45 GHz. It is believed that microwave reactions involve absorption of electromagnetic waves by polar molecules (non-polar molecules are inert to microwaves). When molecules with permanent dipole are subjected to an electric field, they become alligned and as the field oscillates their orientation changes and this rapid reorientation provides intense internal heating. In fact, the main difference between microwave and classical heating is that microwave heating is associated in core and homogeneous heating, whereas in classical heating the heat transfer takes place by preheated molecules. Some of the common solvents that are used in MW ovens include formamide, methanol, ethanol, chlorobenzene, ethylene glycol, dioxane and diglyme. The microwave reactions can be performed in water, organic solvents or even in solid state [For more details about microwaves see V.K. Ahluwalia and R.S. Varma, Alternate Energy processes in chemical synthesis, Narosa Publishing House, 2008, Chapter 1 and 2 and the references cited there in].

Ultrasound is increasingly used for chemical synthesis. The term 'Sonochemistry' is used to describe the effect of ultrasound waves to chemical reactivity. A generally accepted theoretical interpretation is based on the phenomenon of cavitation. The creation of bubbles in a liquid medium, which collapse with the liberation of considerable energy [For more details about ultra sound and its applications in organic synthesis see V.K. Ahluwalia and R.S. Varma, Alternate

Energy processes in chemical synthesis, Narosa Publishing House, 2008, Chapter 6 to 10 and the references cited there in].

Photochemical reactions occur by the absorption of electromagnetic radiations to produce electronically excited molecules with give the product of the reactions. The photochemical reactions are highly stereospecific and have been used for the synthesis of highly strained thermodynamically unstable compounds. The photochemical reactions are used in modern synthetic chemistry and lead to products which are inaccessible by thermal reactions. [For more details about photoinduced reaction see V.K. Ahluwalia and R.S. Varma, Alternate Energy processes in chemical synthesis, Narosa Publishing House, 2008, Chapter 15 and the references cited there in].

5. Use of catalyst

Catalysts are known to facilitate chemical transformations, which are effected in much short time, consuming less energy and giving good yields. An advantage is that the catalysts are not consumed in the reaction and can often be recycled. Even in those case, where no reaction occurs usually, the reaction becomes feasible. An example of this type is the hydrations of alkynes to give aldehydes or ketones.

$$\underset{\text{Acetylene}}{HC\equiv CH} \xrightarrow[H_2SO_4]{HgSO_4} \underset{\text{Acetaldehyde}}{CH_3CHO}$$

$$\underset{\text{Propyne}}{H_3CC\equiv CH} + CO + CH_3OH \xrightarrow{Pd} CH_3-\overset{\overset{}{}}{\underset{\underset{O}{\|}}{C}}-\overset{\overset{O}{\|}}{C}-OCH_3$$

Methyl methacrylate
(Shell corporation)

The catalysts have also been used for selective reduction of a triple bond to a double bond.

$$\underset{\text{Propyne}}{H_3CC\equiv CH + H_2} \xrightarrow{Pd\text{-}BaSO_4} \underset{\text{Propene}}{H_3C-CH=CH_2}$$

Another type of catalysts, which are commonly used for reactions which give low yields (or the reaction does not take place) are the phase transfer catalysts (PTC). The phase transfer catalysts are ionic substances, usually quaternary ammonium salts, where the size of the hydrocarbon group in the cation is large enough to confer solubility of the salt in organic solvents (the cations must be highly lipophilic). In fact, the PTC reactions describe a methology for accelerating reactions between water insoluble organic compounds and water soluble reactants, e.g., reaction of an organic halide with sodium cyanide. The PTC transfers the anion of the reactant from the aqueous phase to the organic phase.

A typical example of a PTC reaction is the conversion of 1-chlorooctane into 1-cyanooctane using a small quantity of an appropriate PTC.

$$CH_3(CH_2)_6CH_2Cl \xrightarrow[CH_3(CH_2)_{15}P^+(n\text{-}Bu)_3 \\ \Delta,\ 105°,\ 2\ hr.]{NaCN,\ H_2O,\ decane} \underset{\text{1-Cyanooctane}}{\underset{95\%}{CH_3(CH_2)_6CH_2CN}}$$

Another example is the oxidation of toluene which $KMnO_4$ in presence of crown ether in much better yields.

$$\text{Toluene } (C_6H_5CH_3) \xrightarrow[\text{Crown ether (Catalytic ammuni)}]{[O] \\ aq{:}KMno_4} \text{Benzoic acid } (C_6H_5COOH)\ > 85\%$$

The crown ethers are a family of cyclic polyethers and like PTC can be used in a number of transformations. A typle example of a crown ethers is dicyclohexyl [18] crown-6. [For more details about PTC and Crown ethers see V.K. Ahluwalia and R. Aggarwal, Organic synthesis, special Techniques, Narosa Publishing House, 2006, Chapter 1 and 2 and the references cited there in].

Besides the points mentioned above, the following be kept in mind for planning a green synthesis.

- The nature of be waste products (or byproducts) should not be toxic or environmentally harmful.
- Whenever possible, the synthetic methodologies should be designed in such a way that the products or byproducts generated should possess no toxicity to human health and environment.
- The raw materials should be renewable rather than depleting, whenever technically and economically practicable.

6. Use of polymer supported substrates or polymer supported reagents.

In the conventional organic synthesis the substrate is treated with a reagent or reagents and the formed product is isolated from the reaction mixture by procedures involving extraction, precipitation, distillation, crystallisation and chromatographic procedure. However, use of polymer supported substrates or polymer supported reagents simplifies the procedure. (N.K. Mathur, C.K. Narang, R.E. Wulliams, Polymers as Aids in organic chemistry, Academic Press, New York, 1980)

Advantages of using polymer supported substrates or reagents

- The reaction can be carried out easily and the isolation of product becomes easier.
- The purification of the product is simplified
- Polymer supported reagents can be recycled after use.
- Most of the reactions can be carried out, cleanly, rapidly and in high yields under mild conditions.
- After isolation of the polymer supported product, the polymer support is cleaved to get the final desired product

There are three types of polymer supported organic synthesis

Type 1 In this type of synthesis, the organic substrate is covalently bonded to the polymer support and then reacted with the reagent, catalyst etc. The formed product remains bounded to the polymer support. The product can be isolated by hydrolysis. Various steps involved are given below (The most widely used polymer is polystyrene).

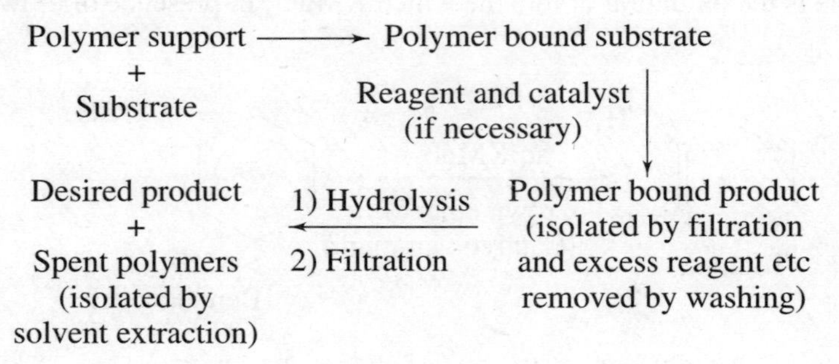

Type 2 In this type of synthesis, the reagent is linked to a polymeric material, which is then reacted with the substrate. After the reaction is over, the polymer support reagent is removed by filtration (and can be reused). The desired product is isolated from the reaction mixtime. Various steps involved are given below.

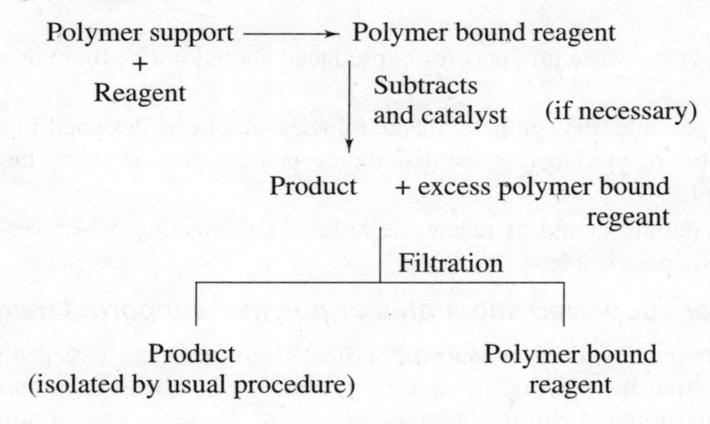

Type 3 This is a polymer supported catalytic reaction. In this type of reaction, conventional catalysts, which are normally used in the homogeneous phase is linked to a polymer support and used in this form to catalyse the reaction.

For more details about polymer supported organic synthesis see V.K. Ahluwalia and R. Aggarwal, Organic synthesis special Techniques, Norosa Publishing House, 2006, Chapter 5. and the reference cited there in.

PART II

GREEN ALTERNATIVES TO SYNTHESIS ORGANIC TRANSFORMATIONS

Most of the organic reactions require the use of volatile organic solvents, anhydrous conditions and generate byproducts which are detrimental to human health and the environment. However, with the development of green chemistry, it is now possible to carry out large number of organic reactions in ecofriendly conditions using newer and better techniques.

In this part, some organic transformations, which are green synthesis and can be easily performed in a chemical laboratory are described. Few examples of each of the following types of transformations are described.

1. Aqueous Phase transformation
2. Transformations in solid phase
3. Photochemical transformations
4. Transformations using Phase Transfer Catalysts
5. Transformations using sonication
6. Transformations using microwave irradiation
7. Enzymatic transformation
8. Transformations in ionic liquids
9. Transformations in polyethylene glycol
10. Transformations involving polysupport reagends or substrates
11. Miscellaneous transformations

Chapter 2

Aqueous Phase Transformations

2.1 p-ACETYLAMINOPHENOL (TYLENOL)

It is an analgesics and is obtained by the acetylation of p-aminophenol with acetic anhydride in aqueous solution.

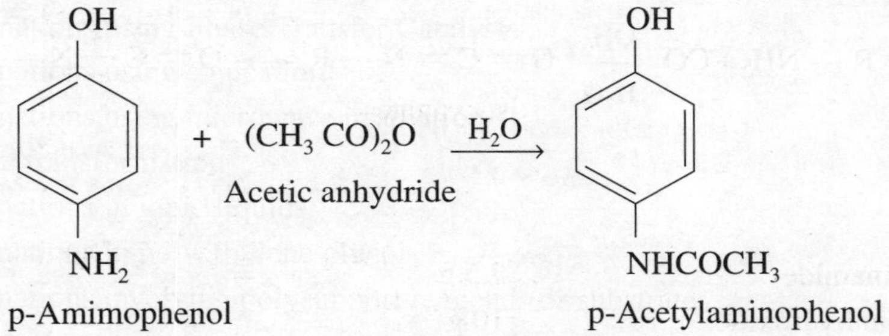

OH

NH$_2$

p-Amimophenol

$$+ \; (CH_3\,CO)_2O \xrightarrow{H_2O}$$

Acetic anhydride

OH

NHCOCH$_3$

p-Acetylaminophenol

Materials

p-Aminophenol	4.4 g
Acetic anhydride	4.8 mL
Water	12 mL

Procedure

Acetic anhydride (4.8 mL) is added to a stirred suspension of p-aminophenol (4.4 g, 0.04 mol) in water (12 mL). The mixture is stirred vigorously and warmed (water bath). The solid dissolves. On cooling, p-acetylaminophenol separates, which is filtered washed with ice-cold water and crystallised form hot water M.P. 169°, yield 5 g (82%).

Notes

1. In p-aminophenol, out of the two groups present, viz., hydroxyl and amino, only the amino group is acetylated in aqueous medium. However, the hydroxyl group can be acetylated in presence of aqueous alkali.
2. In case the final product is not pure (as indicated by its m.p.), it is purified by dissolving in cold dilute alkali and neutralisation of the clear solution by careful addition of dilute hydrochloric acid in cold.

2.2 3-AMINOPYRIDINE

It is obtained Hofmann reaction (Hofmann rearrangement) involving treatment of nicotamide with sodium hypobromite

$$R-\underset{\underset{O}{\|}}{C}-NH_2 \xrightarrow{^{-}OBr} R-\underset{\underset{O}{\|}}{C}-\ddot{N}(Br)H \xrightarrow{KOH} \left[R-\underset{\underset{O}{\|}}{C}-\ddot{N}-Br \right]^{-} \xrightarrow{-Br^{-}} O=\overset{\oplus}{C}-\ddot{N}^{\ominus}-R \longleftrightarrow O=C=N-R \xrightarrow[H_2O]{\bar{O}H} R-NH_2+CO_2$$

R = 3-pyridyl (pyridine ring, N)

isocyanate: $O=C=N-R$

Materials

Nicotinamide	7.5 g
Sodium hydroxide	10 g
Bromine	4 mL

Procedure

Finely powdered nicotinamide (7.5 g 0.061 mol) is added to a solution of sodium hypobromite [prepared by adding bromine (4 mL) in one lot to a cooled solution (< 0.5°) of sodium hydroxide (10 g) in water (10 mL); the mixture is shaken till all the bromine has reacted], the mixture is shaken vigorously for 10-15 min. The solution is warmed to 75° (water-bath) for 40 min, cooled, saturated with NaCl and extracted with ether (3 × 15 mL). The combined etter extract is dried (KOH pellets) and etter removed (distillation on water bath) to give 3-aminopyridine. It is crystallised from benzene-petroleum ether, (4:1), m.p 63° yield (65%).

Notes

1. A.W. Hofmann, Ber. 1881, **14**, 2725.
2. The required nicotinamide is prepared by the esterification of nicotinic acid followed by treatment of the formed ethyl nicotinate with ammonia (V.K. Ahluwalia and Renu Aggarwal, Comprehensive Practical Organic Chemistry, Universities Press, 2004, Page 132).

2.2a Anthranilic Acid

It is obtained by heating phthalalic anhydride with urea followed by treatment of the formed phthalimide with sodium hypobromite.

The last step is **Hofmann Reaction (Hofmann Rearrangement**) (see also preparetion of 3-aminopyridine, Section 2.2)

Phthalic anhydrite $\xrightarrow[\Delta]{\text{Urea}}$ Phthalimide $\xrightarrow{\text{NaOBr}}$ Anthranitic acid

Procedure

An intricate mixture of phthalic anhydride (12 g, 0.084 mol) and urea (2.25 g) is heated in an R.B flask at 130-135° (oil bath). The mixture melts, effervescence commences and frothing takes place (10-15 min). The temperature of the mixture rises to 150-160° and becomes solid. It is cooled, water (20 mL) is added, solid product filtered and washed with water to give **phthalimide**, m.p. 233-234° yield 11 g (88.7%).

Bromine (2.1 mL, 0.041 mol) is added in one lot to a cooled solution (< 0°) of NaOH (7.5 g. in 10 mL H_2O). The mixture is shaken till all the bromine has reacted. The temperature is maintained below 0° and powered phthalimide (6 g, 0.04 moL) added in one lot. The mixture is shaken and a cold solution of NaOH (5.5 g) in water (20 mL) added. The reaction mixture is shaken, its temperature rises (70°). The mixture is warmed to 80° for 5 min. To the cooled solution is added conc. HCl (15 mL) with stirring till the solution is just neutral. Anthranilic acid is precipitated by gradual addition of acetic acid (5-6 mL). The separated product in filtered, washed with water and crystallised from hot water, m.p. 145° yield 3.5 g (63%).

2.3 BENZILIC ACID

It is prepared by the reaction of benzil (an α-diketone) with alkali. The reaction is called **Benzil-Benzilic Acid Rearrangement** (L. Liebig, Ann., 1838, **25**, 27; N. Zinin, Ann., 1939, **31**, 329).

$$C_6H_5-\underset{\|}{\overset{O}{C}}-\underset{\|}{\overset{O}{C}}-C_6H_5 \xrightarrow[KOH]{\overset{\ominus}{OH}} C_6H_5-\overset{O}{\overset{\|}{C}}-\underset{C_6H_5}{\underset{|}{\overset{\bar{O}K^+}{\overset{|}{C}}}}-OH \longrightarrow C_6H_5-\underset{C_6H_5}{\underset{|}{\overset{\bar{O}K^+}{\overset{|}{C}}}}-\overset{O}{\overset{\|}{C}}-OH$$

Benzil
(α-diketone)

$$\xrightarrow{H^+ shift} C_6H_5-\underset{C_6H_5}{\underset{|}{\overset{OH}{\overset{|}{C}}}}-\overset{O}{\overset{\|}{C}}-\overset{-}{O}\overset{+}{K} \xrightarrow{H^+} C_6H_5-\underset{C_6H_5}{\underset{|}{\overset{OH}{\overset{|}{C}}}}-\overset{O}{\overset{\|}{C}}-OH$$

Pot Benzilate

Benzilic acid
(α-hydroxy-acid)

Materials

Benzil	4 g
Potassium hydroxide pellets	5 g (in 10 mL H_2O)
Rectified spirit	12 mL
Cone. Hydrochloric Acid	17 mL

Procedure

A mixture of benzil (4 g, 0.019 mol) potassium hydroxide solution (5 g dissolved in 10 mL water) and rectified spirit (12 mL) is refluxed (water-bath) for 12 min. The hot reaction mixture is poured into water (70 mL). To the mixture, animal charcoal (1 g) is added, mixture stirred and filtered. The clear filtrate is poured with stirring on to a mixture of conc. hydrochloric acid (17 mL) and crushed ice (75 g). The separated benzilic acid is filtered and crystallised from hot water, m.p. 150°, yield 3 g (69%).

Notes

1. In case the reaction does not go to completion, a colloidal solution is obtained due to the separation of unreacted benzil.
2. Benzilic acid can also be obtained by the oxidation of benzoin with alkaline potassium bromate.

$$C_6H_5-CH(OH)\underset{\underset{O}{\|}}{C}C_6H_5 \xrightarrow[KBrO_3]{alk} \left[C_6H_5-\underset{\underset{O}{\|}}{C}-\underset{\underset{O}{\|}}{C}-C_6H_5\right] \longrightarrow (C_6H_5)_2\underset{\underset{OH}{|}}{C}-CO\overset{-}{O}\overset{+}{k}$$

Benzoin

Benzil

Thus to a solution of sodium hydroxide (3.7 g) and potassium bromate (1 g) (or sodium bromate 0.9 g) in water (7 mL) at 85-90° is added benzoin (4 g, 0.019 mol). The temperature is not allowed go above 90°. A small quantity of water is added from time to time to prevent the mixture becoming too thick. The heating is continued at 90° for 1.5-2 hr, until a test portion is almost completely soluble in water. To the reaction mixture is added water (30 mL), resulting mixture cooled (ice-water). In case any byproduct consisting of benzhydrol separates, it is filtered and the clear filtrate is acidified (dil. H_2SO_7, 1:3, 13 mL). The separated benzilic acid is filtered and crystalized from water m.p. 149-50° yield 3.1 g (72%)

In the above preparation, in place of benzoin, benzil can also be used.

4. Benzil-Benzilic acid rearrangement proceeds efficiently and faster is solid state by MW irradiation reducing considerably the time of the reaction and better yields (F. Toda, Y. Kagawa and Y. Sakaino, Chem, Lett., 1990, 373).

$$\mathrm{Ar-\overset{O}{\overset{\|}{C}}-\overset{O}{\overset{\|}{C}}-Ar'} \xrightarrow[\text{3-5 min}]{\text{KOH/MW}} \mathrm{Ar-\underset{OH}{\overset{Ar'}{C}}-COOH}$$

>95%

2.4 BENZOIN

The self-condensation of aromatic aldehydes (with no α-hydrogen) in presence of cyanide ions as catalyst to give α-hydroxy ketone (benzoin) is called **benzoin condensation**. Thus benzaldehyde on treatment with potassium cyanide in rectified spirit give benzoin.

$$\mathrm{C_6H_5-\overset{O}{\overset{\|}{C}}-H} + :\mathrm{CN}^{\ominus} \rightleftharpoons \mathrm{C_6H_5-\underset{CN}{\overset{O^{\ominus}}{C}}-H} \rightleftharpoons \mathrm{C_6H_5-\underset{CN}{\overset{OH}{C^{\ominus}}}} + \mathrm{H-\overset{O}{\overset{\|}{C}}-C_6H_5} \rightleftharpoons$$

Benzaldehyde

$$\mathrm{C_6H_5-\underset{CN}{\overset{O-H}{C}}-\underset{H}{\overset{O^{\ominus}}{C}}-C_6H_5} \rightleftharpoons \mathrm{C_6H_5\overset{O}{\overset{\|}{C}}-\underset{H}{\overset{OH}{C}}-C_6H_5}$$

Benzoin
(α-Hydroxyketone)

Materials

Benzaldehyde	3.6 mL
Rectified spirit	8 mL
Potassium cyanide	0.5 g (in 3 mL H_2O)

Procedure

A mixture of benzaldehyde (3.6 mL, 0.035 mol) potassium cyanide solution (0.5 g in 3 mL water) and ractified spirit (8 mL) is gently refluxed for 30 min is a 100 mL R B flask. The reaction mixture is cooled (ice-bath), separated product filtered, washed with water and crystallised from rectified spirit to give benzoin, m.p. 137° yield 3.2 g (42.7%).

Notes

1. A.J. Lapworth, J. Chem. Soc., 1903, **83**, 955; 1904, **85**, 1206; J. Solodar, Tet. Lett., 1971, 287
2. Potassium cyanide is deadly poisonous. It should be handeled with extreme care using hand gloves. The filtrate should be disposed off into sink and washed with large quantity of water.
3. Benzoin condensation can only be performed with aromatic aaldehydes (with no α-hydrogen. The reaction is not successful with aliphatic aldehydes under these conditions.
4. It is found that benzoin condensations of aldehydes are strongly catalysed by quaternary ammonium cyanide in a two phase system (J. Solodar, Tet. Lett., 1971, 287).
5. Benzoin condensation of anisaldehyde and p-tolualdehyde give the corresponding α-hydroxy ketone. Similarly furfural gives furoin.

2.5 BENZOTRIAZOLE

It is prepared by the reaction of o-diaminobenzene with nitrous acid (generated in situ from sodium nitrite and acetic acid)

NH₂ NH₂ + N(=O)OH → N(H)–N(OH)–OH / NH₂ → N=N–OH / NH₂

O-Diamino benzene Nitrous acid

$\xrightarrow{-\bar{O}H}$ N≡N⁺ / NH₂ $\xrightarrow{-H^+}$

Benzotriazole

Materials

o-Diaminobenzene	5.4 g
Acetic acid (glacial)	5.7 mL
Sodium nitrite solution in water	3.8 dissolved in 8 mL water

Procedure

A solution of o-diaminobenzene (5.4 g, 0.05 mol) is dissolved in a mixture of glacial acetic acid (5.7 ml, 0.1 mol) and water (15 mL) by slight warming if necessary. To the cooled solution (0°, using ice-bath) is added in one lot a solution of sodium nitrite in water (3.8 g in 8 mL water). The mixture is stirred, temperature rises to 80° and stirring continued for 15 min. The resulting reaction mixture is cooled (ice-water) separated benzotriazole filtered, washed with ice-cold water and crystallised from hot water m.p. 99-100° yield 4 g (67%)

Notes

1. The required o-diaminobenzene is prepared as given in the preparation of Benzimidazole (Section 7.4)

2.6 2-BENZOYL-3,5-DIMETHYLBENZOFURAN

It is obtained by the reaction of 2-hydroxy-5-methylacetophenone with phenacyl bromide in aqueous potassium carbonate in presence of tetrabutylammonium hydrogen sulphate.

OH, H_3C, $COOH_3$ + $COCH_2Br$ $\xrightarrow[\text{aq. } KCO_3]{n\text{-}Bu_4\overset{+}{N}\,HSO_4^-}$ O, CO, H_3C, CH_3

2-Hydroxy-5-methylacetophenone

Phenacyl bromide

70%

2-Benzoyl-3,5-dimethylbenzofuran

Procedure

Stir vigornerly a mixture of 2-hydroxy-5-methylacetophenone (0.01 mol), phenacyl bromide (0.01 mol), tetrabutylammonium hydrogen sulphate (0.005 mol), dichloromethane (110 mL) and aqueous potassium carbonate (30%, 3 mL) for 3 hr at 30°. Separate the organic layer, wash with water, dry ($Na_2\ SO_4$) and evaporate the solvent. Crystallise the residue from ethanol, M.P 152°, yield 70%.

Notes

1. G. Sabitha and A.V. Subba Rao, Synth. Commun, 1987, **17**, 341
2. The PTC, tetrabutylammonium hydrogen sulphate is prepared as described in Section 5.8, Note 3.

2.7 n-BUTYL BROMIDE

It is obtained by the reaction of n-butyl alcohol and sodium bromide solution with conc. H_2SO_4.

$$\underset{\text{n-Butly alcohol}}{CH_3CH_2CH_2CH_2OH} + \underset{\text{(aq)}}{NaBr} \xrightarrow{\text{Conc. } H_2SO_4} \underset{\text{n-Butylbromide}}{CH_3CH_2CH_2CH_2Br}$$

Procedure

n-Butyl alcohol (12 mL) is added to a solution of sodium bromide (16.2 g) in water (18 mL) contained in a R.B. flask (100 mL capacity). Conc. H_2SO_4 (13.7 ml) is added to the above solution slowly with constant shaking (temp. is kept below 40°). The mixture is refluxed for 1 hr. The reaction mixture is cooled, extracted with ether and ether distilled to give n-butyl bromide b.p. 101-102°. yield 12 g (66.8%)

Notes

1. n-Butyl bromide can also be obtained by the reaction of n-butyl alcohol with HBr (48%) in presence of conc. H_2SO_4.
2. Isopropyl bromide (b.p. 59°) is obtained by heating isopropyl alcohol with hydrobromic acid (48%)

2.8 TERT.BUTYLCHLORIDE

It is obtained by treating tert. butanol with hydrochloric acid

$$\underset{\text{tert. Butanol}}{(CH_3)_3C - OH + HCl} \longrightarrow \underset{\text{tert Butylchloride}}{(CH_3)_3C - Cl + H_2O}$$

Procedure

A mixture of tert butyl alcohol (15 g) and conc. HCL (50 mL) is stirred for 20-25 mm. The lower layer of tert. butylchloride is separated (using a separatory funnel), washed with $NaHCO_3$ solution, water, dried ($CaCl_2$) and distilled. Yield 17 g (88-89%) B.P. 49-50°.

Notes

1. n-Primary alcohols be converted into the corresponding chloride by refluxing with thionyl chloride. Using this procedure, n-hexylchloride, b.p. 132-133° is obtained in about 35% yield.
2. Secondary alcohols can be converted into the corresponding chlorides by refluxing the alcohol with thionly chloride in presence of pyridine. Using this procedure isobutylchloride, b.p. 68-69° is obtained in about 70% yield.

2.9 CHALCONE (BENZALACETOPHENONE)

It is obtained by the condensation of benzaldehyde with acetophenone in presence of alkali. It is known as **Claisen-Schmidt Condensation**. (L. Claisen and A. Claparede, Ber., 1981, **14**, 2460, J.B. Schmidt, Ber. 1881, **14**, 1459)

$$\underset{\text{Acetophenone}}{C_6H_5-\overset{\overset{\displaystyle O}{\|}}{C}-\underset{\underset{\displaystyle H}{|}}{CH_2}} + {}^{\ominus}OH \rightleftharpoons C_6H_5-\overset{\overset{\displaystyle O}{\|}}{C}-\overset{\ominus}{C}H_2 + \underset{\text{Benzaldehyde}}{C_6H_5\overset{\overset{\displaystyle O}{\|}}{C}-H}$$

$$\rightleftharpoons C_6H_5-\overset{\overset{\displaystyle O}{\|}}{C}-CH_2-\overset{\overset{\displaystyle O^-}{|}}{C}H-C_6H_5 \xrightarrow{-H_2O} \underset{\substack{\text{Chalcone}\\ \text{(Benzalacetophenone)}\\ (\alpha,\beta\text{-unsatd ketone})}}{C_6H_5-\overset{\overset{\displaystyle O}{\|}}{C}-CH=CH\,C_6H_5}$$

Materials

Acetophenone	5 mL
Benzaldehyde	4.4 mL
Rectified spirit	15 mL
Sodium hydroxide solution	2.2 g. in 20 mL water

Procedure

To a cooled and stirred solution (ice-bath) of sodium hydroxide (2.2 g) in water (20 mL) and rectified spirit (15 mL) is added acetophenone (5 mL, 0.043 mol) followed by the addition of benzaldehyde (4.4 mL, 0.043 mol). The reaction mixt. is kept at 25°, stirred for 2-2.5 hr. until the mixture becomes viscous and no further stirring is possible. The reaction mixture is kept over night in a refrigerator, separated product filtered, washed with water and recrystallised from rectified spirit, m.p. 56-57° yield 7.7g (86.5%).

Notes

1. Benzalacetophene (Chalcone) is skin irritant and should be handeled with care.
2. Similar condensation of acetone with benzaldehyde (in equimolar amounts) gives **benzalacetone** ($(C_6H_5CH=CH-\overset{\overset{\displaystyle O}{\|}}{C}-CH_3)$, m.p 42° (b.p. 143-144°/20 mm) in 85%

yield. However use of double the amount of benzaldehyde in the above condensation gives **dibenzalacetone** ($C_6H_5CH = CH—\overset{O}{\overset{\|}{C}}—CH = CH—C_6H_5$), m.p. 112° in 84.7% yield.

3. Chalcones have also be obtained by the condensation of equimolar amount aromatic acedehydes with ketones in presence of a catalytic quantity sodium hydroxide (1-2 pellets) by heating in a microwave oven for 30 sec. to 2 min (R. Gupta A.G. Gupta, S. Paul and P.L. Kachroo, India J. Chem. 1995, **34B**, 61)
4. Phase transfer catalysts such as cetyltrimethyl ammonium compounds (CTACl or CTABr etc.) have been successfully used for claisen smith reaction of acetophenone with benzaldehydes [K.R. Nivalkar, C.D. Mudaliar and S.H. Mashraqui, J. Chem. Res. (S), 1992, 98; F.Fringulli, G. Pani, O. Piermatti and F. Pizzo tetrahedron, 1994, **50**, 11499; F. Fringulli, G. Pani, O. Piermatti and F. Pizzo, tetrahedron, 1994, **50**, 114 88; F. Fringulli, G. Pani, O. Piermatti and F. Pizzo Life Chem. Rep; 1995, **13**, 133].

R—(C₆H₄)—$COCH_3$ + ArCHO $\xrightarrow[\text{RT, 0.3-16 hr; 61-94\%}]{\text{PTC, NAOH}}$ R—(C₆H₄)—CO—CH=CH—AR

Acetophenones — Chalcones

2.10 CYCLOHEPTANONE

It is obtained from cyclohexanone by reaction with diazomethane generated in situ from N-methyl-N-nitrosotoluene-p-sulphonamide and alkali. The reaction involved ring expansion and is represented as:

Cyclohexanane $\xrightarrow{CH_2N_2}$ Cycloheptenane (60%) + [Epoxide + Cyclooctanone]

The mechanism of the reaction is given below.

Cyclohexanane + $:\overset{\ominus}{C}H_2 - \overset{\oplus}{N}\equiv N$ (Diazo methane) ⟶ intermediate; route a ⟶ Cycloheptanone; route (b) ⟶ epoxide; Cycloheptanone $\xrightarrow{CH_2N_2}$ Cyclootanone

Materials

Cyclohexanone	0.49 g
N-Methyl-N-nitrosotoluene-p-sulphonamide	12.5 g
Potassium hydroxide solution	1.5 g in 5 mL aq. EtOH (1 : 1)

Procedure

To a stirred and cooled (0°) mixture of cycyclohexanone (0.49 g, 0.05 mol) and N-methyl-N-nitrosotoluene-p-sulphonamide (12.5 g, 0.0585 mol) in ethyl-alcohol-water (15 mL 95% EtOH and 1 mL water) is added dropwise a solution of KOH (1.5 g) in aqueous ethanol (5 mL 1:1). A vigorous reaction starts and temperature rises. In case the reaction does not start, remove the flask from ice-bath and warm the mixture to 10°. No further addition of alkali is made till the reaction has started. During addition of KOH solution, the reaction is maintained at 10-20°. Total time of addition is about 1½ to 2 hr. Stirr for 30 minutes more and then acidify with 2 N HCl until the solution is acidic to litmus.

To the remaining solution, add sodium metasulphide solution (10 g in 20 mL water) and stir (10 min.). The separated bisulphite compound is filtered and decomposed with 12.5 g Na_2CO_3 in 15 mL H_2O. Extract the separated cycloheptanone with ether (3 × 5 mL), dry the ether extract (Na_2 SO_4) and distil. Cycloheptanone. b.p. 64-65°/12 mm is obtained in 31% yield (1.7 g)

Note

1. Only the cycloheptanone forms crystalline sodium bisulphite addition product and so is easy to separate from the byproduct (cyclooctanone and the epoxide)

2. Diazomethane is generated in situ as show below

$$p\text{—}CH_3.C_6H_4.SO_2\,N(NO)CH_3 + KOH \rightarrow p.CH_3.C_6H_4.SO_3K + CH_2N_2 + H_2O$$

N-Methyl-N-Nitrosotoluene
-p-sulphonamide

2.11 2,3-DIHYDROXY ANISOLE (PYRIGALLOL MONOMETHYL-ETHER)

It is usually prepared by **Dakins oxidation**[1]. The reaction involved oxidation of o-hydroxy aldehydes with hydrogen peroxide in presence of aqueous sodium hydroxide. The mechanism of the oxidation is given below.

OH, C—H, O, $HO\overset{\ominus}{O}$ $\xrightarrow[\text{NAOH}]{H_2O_2}$ O—H, CH, $\ddot{O}$—O—H, $O^{\ominus}$ $\xrightarrow[-H^+]{-OH^-}$ $O^{\ominus}$, O—C—H, O

$\xrightarrow{^-OH}$ OH, OH + HCOOH

Catechol

2.3-Dihydroxy anisole is obtained by the oxidation of 2-hydroxy-3-methoxybenzaldehyde (o-vanillin) with hydrogen peroxide in presence of aqueous sodium hydroxide

OCH_3, OH, CHO $\xrightarrow[H_2O_2/\text{aq. NaOH}]{[O]}$ OCH_3, OH, OH

o-Vanillin 2,3-Dihydroxyanisole

Materials

o-Vanillin	5.1 g
Aqueous sodium hydroxide (2N)	16.7 mL
Hydrogen peroxide (6%)	24 mL

Procedure

To solution of o-vanillin (5.1 g, 0.028 mol) in aqueous sodium hydroxide (2 N, 16.7 mL, 0.028 mol) is added dropwise hydrogen peroxide (6%, 24 mL, 0.42 mol) during 1 hr. at 40-50° (the solution is cooled if necessary). After the reaction is over, the solution is cooled to room temperature, saturated with sodium chloride and extracted with ether (3 × 25 mL). The combined ether extract is dried (sodium sulphate) and distilled. The residual product is distilled and fraction boiling at 136-38°/22 mm is collected. The yield is 2.8 g (68%)

Notes

1. H.D. Dakin, 0.S, 1941, I, 149; J.L. Letter, Chem. Rev., 1949, **45**, 385
2. Using this procedure, olter o-hydroxyaldehydes like salicyaldehyde can also be oxidised to the corresponding hydroxy compound (CHO → OH).
3. Normally, the yields in Dakins oxidation are low. This reaction has now been carried out in high yields using sodium percarbonate (SPC, Na_2 CO_3, 1.5 H_2O) in H_2O-THF under ultrasonic irradiation (G.W. Kablka, N.K. Reddy and C. Narayana, Tetrahedron Lett., 1992, **33**, 865). Using this procedure a number of aldehydes have been oxidised in 85-90% yield. The aldehydes oxidised include o-hydroxybenzaldehyde, p-hydroxybenzaldehyde, 2-hydroxy-4-methoxybenzaldehyde, 2-hydroxy-3-methoxybenzaldehyde and 3-methoxy-4-hydroxybenzaldehyde.
4. Dakins oxidation can also be performed in solid state by heating together an o-hydroxyzaldehydes and urea-formaldehyde complex at 50 to 80° in good yield (80-85%). This solid state oxidation forms the subject matter of a subsequent section (for details Sec. 3.9)
5. Aldehydes can also be converted into hydroxyl group by oxidation with peracids like peracetic acid. In such oxidation it is not necessary to have an hydroxyl group in ortho position. Such oxidation form the subjection matter of a subsequent section (for details Sec. 3.19). This oxidation is known as **Baeyer-Villiger Oxidation**.

2.12 2,4-DIHYDROXYBENZOIC ACID (β-RESORCYLIC ACID)

It is prepared from resorcinol, potassium bicarbonate and carbon dioxide gas. The reaction is known as **Kolbe-Schmitt Reaction**[1]

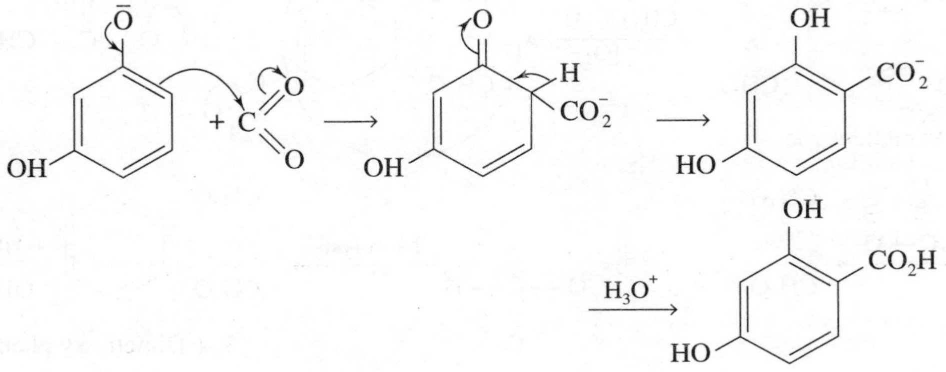

Materials

Resorcinol	15 g
Pol. bicarbonate	75 g
Water	150 mL

Procedure[2]

A mixture of resorcinol (15 g), potassium bicarbonate (75 g) and water (150 mL) is heated in a three necked R.B flask (250 mL capacity) fitted with a gas inlet tube and a reflux condenser. The reaction mixture is heated on a steam bath for 3 hrs and them refluxed for 30 min while passing a rapid stream of CO_2 gas through the solution. The hot solution is acidified[3] (conc. HCl, 45 mL). The reaction mixture is cooled (ice-bath) and the separated β-resorcylic acid filtered. Yield 9 g (64%). M.P. 216-17°.

Notes

1. H. Kolbe, Ann., 1860, **113**, 125; R. Schmidt, J.prak.Chem., 1885, **(2) 31**, 397; A.S. Lindsey and H. Ieskey, Chem. Rev., 1957, **57**, 583.
2. Nierensltein, Clibbens, Org. Synth. Coll. Vol. II, 1943, 557
3. For acidification conc. HCl is added slowly through a separatory funnel, with a long tube delivering the acid to the bottom of the flask.
4. Using phenol, phloroglucinol or pyrogallol in place of resorcinol in the above reaction, salicylic acid, **2,3,4-trihydroxy benzoic acid** or **2,4,6-trihydroxybenzoic acid** can be prepared.
5. 2-Hydroxy-4-methoxy benzoic acid, m.p 160-161° can be obtained by the oxidation of 2-hydroxy-4-methoxyacetophenone with iodine-pyridine (V.K. Ahluwalia and R. Aggarwal, Comprehensive practical organic chemistry, Universities Press, 2004, P 173)

2.13 3, 4-DIMETHOXYPHENOL

It is prepared by the **Baeyer-villiger oxidation** of veratraldehyde with peracetic acid

CH_3O, CH_3O, CHO

Verataldehyde

$\xrightarrow[\text{[O]}]{CH_3CO_3H}$

CH_3O, CH_3O, C, O—H, O—O—C(=O)—CH_3

$\xrightarrow{-CH_3C(=O)-O^{\ominus}}$

CH_3O, CH_3O, O—C(=O)—H

$\xrightarrow{\text{hydrolysis}}$

CH_3O, CH_3O, OH + HCOOH

3, 4-Dimethoxy phenol

Materials

Verataldehyde	5 g
Acetic acid (glacial)	30 mL
Peracetic acid solution	15 mL
Potassium hydroxide	10 g

Procedure

To a solution of verataldehyde (5 g, 0.03 mol) in glacial acetic acid (30 mL) is added dropwise a solution of peracetic acid (15 mL) durring 30 min. The reaction mixture is left for about 10 hrs. and then concentrated in vacuo to about 5 mL. The residue is extracted with ether (2 × 15 mL). The ether is remove by distillation and the residual oily product is hydrolysed by refluxing with potassium hydroxide (10 g) in aqueous alcohol (1 : 4, 100 mL) for 1 hr. The reaction mixture is concentrated in vacuo almost to dryness. It is dissolved in water and the solution rendered acidic with dilute H_2SO_4. The solution is extracted with ether, ether extract dried (Na_2SO_4) and distilled. The residual product is purified by column chromatography over silica gel (2 × 30 cm column). Elution with benzene gave 3, 4-dimethoxyphenol. It crystallised from benzene, m.p 78-80°. Yield 3g (65.2%).

Notes

1. Peracetic acid is prepared in situ by adding 1 part of H_2O_2 (30%) to glacial acetic acid (3 parts) in presence of catalytic amount of H_2SO_4.
2. Baeyer-Villeger oxidation of ketones give esters ($C_6H_5\ COCH_3 \xrightarrow{PhCO_3H} C_6\ H_5OCOCH_3$) (A.V. Balyer and V. Villiger, Ber. 1899, **32**, 3625.

2.14 2,3-DIMETHYL-1-PHENYLPYRAZOL-5-ONE

Also known as **antipyrene**, it was used as antipyretic analgesic for a long time. However, due to its slight action on heart, it has been replaced by aspirin. It is prepared by the methylation of 3-methyl-1-phenylpyrazol-5-one by dimethyl sulphate in aqueous alkaline solution.

H_3C, N, N, O, C_6H_5 $\xrightarrow[\text{aq. NaOH}]{(CH_3)_2\ SO_4}$ H_3C, N, H_3C, N, O, C_6H_5

3-Methyl-1-phenyl pyrazol-5-one

2,3-Dimethyl-1-phenyl pyrazol-5-one (antipyrin)

Materials

3-Methyl-1-phenyl pyrazol-5-one	4.35 g
Sodium hydroxide solution	1 g in minimum amount of water
Dimethyl sulphate	3.6 g, (2.7 mL)

Procedure

Sodium hydroxide solution (1 g dissolved in minimum amount of water) is added to a stirred solution of 3-methyl-1-phenylphrazol-5-one (4.35 g, 0.025 mol) in methanol (5mL). The mixture is refluxed (water bath) for about 5 min. Methanol is distilled off and hot water added to the residual mixture. The clear solution is extracted with benzene (3 × 20 mL). The benzene extract is evaporated and the residual product crystallised from hot water (charcoal). yield 3.5 g (74%) M.P. 113°.

Note

1. The required 3-methyl-1-phenylpyrazol-5-one is prepared as given in Section 7.13.
2. The above methylation (N-alkylation) can be conventiently effected under sonication in the presence of a phase transfer catalyst (polyethylene glycol monomethyl ether) in 60% yield. (R.S. Davidson, A.M. Patel, A. Safdar and D. Thornthwaite, Tetrahedran Lett, 1983, **24**, 5907)

H_3C, N, N, O, C_6H_5 —— MeI/Solid KOH/toluene / PEG methyl ether / 20°, 30 min.))) ——→ H_3C, N, Me, N, O, C_6H_5

In place of PTC, crown ether can also be used (J. Jurczak and R. Ostaszewki, Tetrahedron, Lett. 1988, **29**, 959)

2.15 3, 5-DIMETHYLPYRAZOLE

It is prepared by stirring together pentane-2, 4-dione and hydrazine (generated by the action of alkali on hydrazine sulphate)

$CH_2 \cdot CO \cdot CH_3$ / $CH_3 \cdot CO$ (Pentane 2, 4-dione) —— $NH_2 \cdot NH_2$ ——→ CH_3, H_3C, N, H (3,5 Dimethylpyrazole)

Materials

Pentane-2, 4-dione	5 g
Hydrazine sulphate	6.5 g
Sodium hydroxide solution	2.5 M 40 mL

Procedure

Pentane-2, 4-dione (5 g, 5.15 mL 0.05 mole) is added to a stirred mixture of hydrazine sulphate (6.5 g) in sodium hydroxide solution (2.5 M, 40 mL in water). The temperature is maintained at about 15°. After the reaction is over (30 min. stirring), water (20 mL) is added, mixture extracted with ether (20 mL), ether extract washed with water, dried (anhyd. K_2CO_3) and distilled. The separated solid is crystallised from light petroleum, m.p. 107-108° yield 3.6 g (75%)

2.16 5,5-DIPHENYLHYDANTOIN

It is obtained by the condensation of benzil with urea in aqueous alkaline medium. The formed intermediate heterocyclic pinacol on acidification yields the required. 5,5-diphenylhydantoin as a result of **pinacolic rearrangement.**

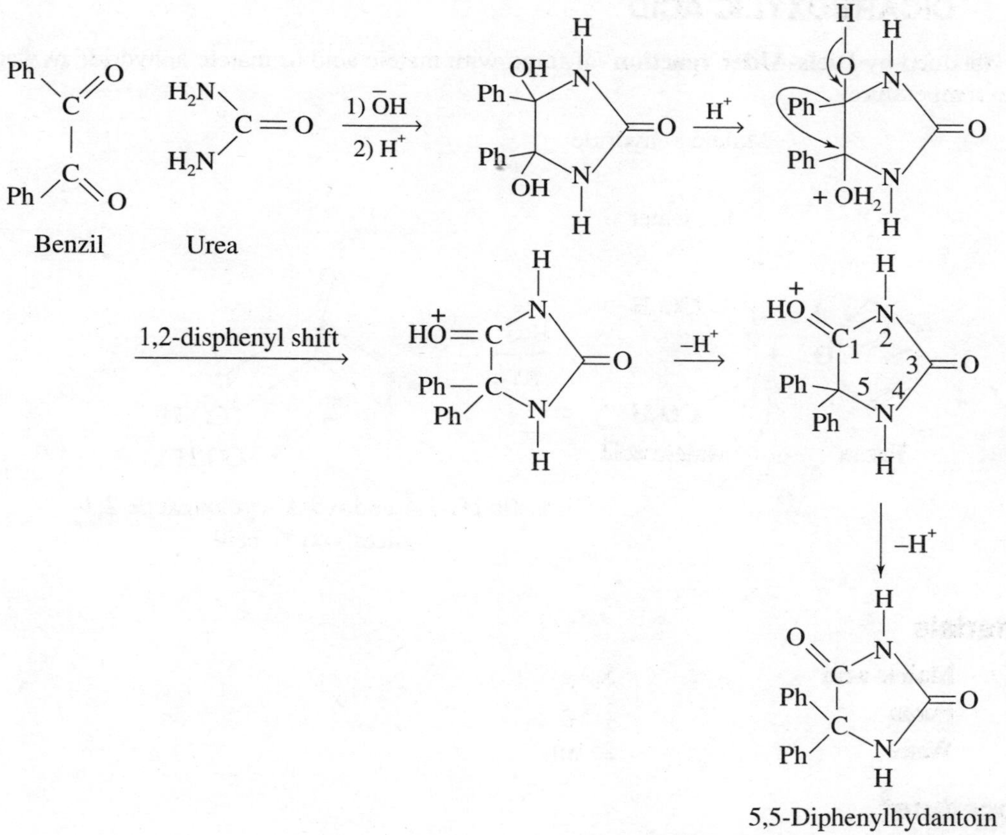

Benzil Urea 1) $\bar{O}H$ 2) H^+ H^+

1,2-disphenyl shift $-H^+$ $-H^+$

5,5-Diphenylhydantoin

Materials

Benzil	5.3 g
Urea	3 g
Aqueous NaOH solution (30% solution)	15 mL

Procedure

A mixture of benzil (5.3 g, 0.025 mol), aqueous sodium hydroxide (30%, 15 mL) and ethanol (75 mL) is refluxed for 2 hr. The solution is cooled, water (125 mL) added and mixture stirred. The separated byproduct is filtered and the filtrate acidified with conc. HCL and cooled. The formed 5, 5-dipheylhydantion is filtered and crystallised from rectified spirit, m.p 297-298°. yield 2.8 g (44%).

Note

1. The reaction can also be conducted in a microwave oven (3 min. heating) and worked up as described above.

2.17 ENDO-CIS-1,4-ENDOXO-Δ^5-CYCLOHEXENE-2,3-DICARBOXYLIC ACID

It is obtained by **Diels-Alder reaction**[1] of furan with maleic acid or maleic anhydride in water at room temperature.

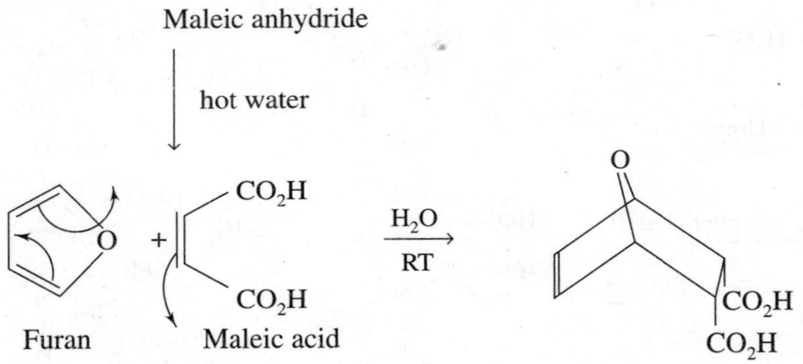

Maleic anhydride

hot water

Furan + Maleic acid (CO_2H, CO_2H) $\xrightarrow[RT]{H_2O}$ (CO_2H, CO_2H)

Endo-cis-1,4-endoxo-Δ^5-cyclohexene-2,3-dicarboxylic acid

Materials

Maleic acid	5.0 g
Furan	2.5 g
Water	25 mL

Procedure[2]

Maleic acid (5.g) and furan (2.5 g) in water (25 mL) is stirred at room temperature for 2-3 hr. The separated product is filtered and washed with water. Yield 7.2 g. Record its m.p

Notes

1. O.Diels and K.Alder, Ann., 1928, **460**, 98; 1929, **470**, 62; Ber., 1929, **62**, 2087; J.A. Norton, Chem. Rev., 1942, **31**, 319.

2. R.B. Woodward and H. Baer, J.Am. Chem. Soc., 1948, **70**, 1161; R. Breslow and D. Rideout, J. Am. Chem. Soc., 1980, **102**, 7816.
3. Diels-Alder reaction proceeds very satisfactory under microwave irradiation. This procedure reduces the time for completion of the reaction to 90 seconds. See preparation of anthracene-malic anhydride adduct (Section 7.1)
4. Diels-Alder reaction proceeds in high yield and high stereo specificy by using ionic liquids like [b min] [BF_4], [b min] [ClO_4], [e min] [CF_3 SO_3] etc. as solvents (T. Fischer, T. Sethi, T welton, J. Woolf, Tetrahedron Lett., 1999, **40**, 793; M.J. Earle, P.B. MeMormal and K.R. Sedden, Green chem., 1999 **1**, 23.

2.18 p-ETHOXYACETANILIDE (PHENACETIN)

It is an analgesic drug and is obtained by the reaction of p-accetamidophenol (Tylenol) with ethyl bromide.

$NHCOCH_3$ $\xrightarrow[pH = 14]{CH_3CH_2Br}$ $NHCOCH_3$

HO $CH_3 — H_2C — O$

p-Acetamidophenol   p-Ethoxyacetanilide (phenacetin)

Materials

p-Acetamidophenol (Tylenol)	1.5 g
Methanol	10 mL
Sodium hydroxide solution (50%)	0.63 mL
Ethylbromide	1.5 mL

Procedure

Sodium hydroxide solution (50%, 0.63 mL) is added to a mixture of p-acetamidophenol (1.50 g) and methanol (10 mL). The mixture is stirred to gel a clear solution. Ethyl bromide (1.5 mL) is added to the above solution and mixture refluxed for 2 hrs. Hot water (20 mL) is added and the mixture cooled. The separated p-ethoxyacetanilide is filtered and crystallised from dilute alcohol. yield 3g (80 %), m.p. 137°.

Notes

1. The required p-acetamidophenol is prepared as described in Section 2.1.

2.19 6-ETHOXYCARBONYL-3,5-DIPHENYL-2-CYCLOHEXENONE

It is obtained by the reaction of chalcone with ethyl acetoacetate in presence of sodium hydroxide. This is known as **Michael addition reaction**. The Michael adduct on base catalysed **aldol condensation reaction** gives 6-ethoxycarbonyl-3, 5-diphenyl-3-hydroxy-cyclohexanone. Final dehydration gives the required 6-ethoxycarbonyl-3,5-diphenyl-2-cyclohexanone. Various steps involved are gives below:

Ethyl acetoacetate + Chalcone $\xrightarrow[\text{(Michael addition)}]{\text{NaOH}}$ Michael adduct $\xrightarrow[\text{aldol condensation}]{\text{NaOH}}$

aldol product, 6-Ethoxycarbonyl-3,5-diphenyl-3-hydroxy-cyclohexanone $\xrightarrow[-H_2O]{\text{NaOH}}$ 6-Ethoxycarbonyl-3,5-diphenyl-2-cyclohexenone

Materials Required

trans-Chalcone	0.72 g
Athyl acetoacetate	0.45 g
Absolute ethanol	15 mL
Sodium hydroxide solution (2.2 M)	0.75 mL

Procedure

Ethyl acetoacetate (0.45 g) and absolute alcohol (15 mL) is added to trans-chalcone (0.72 g) (taken in a R.B. flask). The reaction mixture is shaken to get a clear solution. Sodium hydroxide solution (2.2 M. 0.75 mL) is added to the reaction mixture. The mixture starts boiling (exothermic reaction)

and refluxing continued for 1 hr. The reaction mixture on cooling (ice-bath) gives the required 6-ethoxycarbonyl-3, 5-diphenyl-2- cyclohexenone, which is filtered and crystallised from ethanol. Yield 0.6 g, m.p 111-112°.

Notes

1. A. Garcia Rusto, J. Garcia-Raso, J.V. Sinisterra and R. Mestres, J. Chem. Education, 1986, 63, 443.
2. The base abstracts a proton from the CH_2 group of ethyl acetoacetate and the resulting carbanion attacks the carbonyl group to give a stable 6-membered ring. In this case ethanol supplies a proton to give the aldol intermediate product.
3. The Michael addition of active nitrites to acetylenes can be catalysed by the addition of quaternary ammonium chloride (M. Makosaza, Tetrahedron Lett., 1966, 5489, Polish Patent 55113 (1968), CA, 1969, 70, 106006.

$$C_6H_5\underset{\displaystyle R}{\underset{|}{C}}H—CN + HC \equiv CR' \xrightarrow[\text{DMSO, Solid NaOH}]{C_6H_5CH_2N^+ Et_3Cl^-} C_6H_5—\overset{\displaystyle CN}{\overset{|}{\underset{\displaystyle R}{\underset{|}{C}}}}—CH = CHR'$$

4. Michael addition can also be effect in aqueous medium (Z. Hajos and D.R. Parrisn, J. Org. Chem. 1974, **39**, 1612, U. Elder, G. Sauer and R. Wiechert, Angew. Chem. Int. Edn. Engl., 1971, **10**, 496). An example is given below

2-Methyl cyclopentane 1,3 dione + Methyl vinyl ketone $\xrightarrow[pH>7]{H_2O}$ [intermediate] →

5. Michael addition reactions can also be performed in ionic liquid in presence of a catalyst $Ni(acac)_2$ (M.H. Dell' Anna, V. Gallo, P. Mastrorilli, C. Francesco, G. Rommanazzi and G.P. Suranna, Chem. Commun, 2002, 434). One example is gives below.

$$+ \xrightarrow[[b_{min}]\,[BF_4]]{Ni\,(acac)_2}$$

2.20 HETERO DIELS-ALDER ADDUCT

The nitrogen containing heterocyclic compound can be synthesised using **hetero Diels-Alder reaction**. A typical reaction[1] is the reaction of cyclopentadiene with iminium salt generated from benzylamine hydrochloride and formaldehyde using water as solvent and give **aza-Diel-Alder reaction** products.

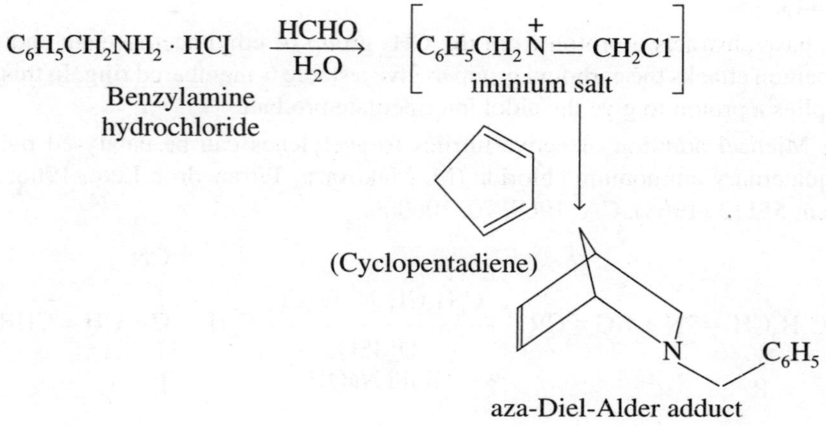

$$C_6H_5CH_2NH_2 \cdot HCl \xrightarrow[H_2O]{HCHO} \left[C_6H_5CH_2\overset{+}{N}=CH_2Cl^-\right]$$

Benzylamine hydrochloride — iminium salt

(Cyclopentadiene)

aza-Diel-Alder adduct

Materials

Cyclopentadiene	13 g
Benzylamine Hydrochloride	11.7 g
Formalin 40%	25 mL

Procedure[1]

Cyclopentadiene (13 g) is added slowly to a stirred mixture of benzylamine hydrochloride (11.7 g) and formalin (25 mL). The stirring is continued for 3 hr. at room temperature. The separated adduct is obtained in quantitative yield (bicyclic amine)

Note

1. S.C. Larnsen and P.A. Griew. J. Am. Chem. Soc., 1985, **107**, 1768
2. The structure of the adduct is confirmed by its NMR spectra.
3. This procedures can be used[1] for the synthesis of large number of bicyclic compounds.

2.21 HIPPURIC ACID (BENZOYL GLYCINE)

It is obtained by the reaction of glycine with benzoyl chloride in presence of sodium hydroxide solution. The reaction is known as **Schotten-Baumann Reaction**.

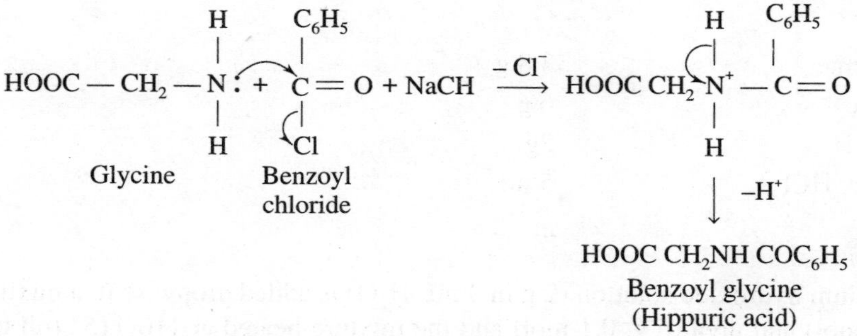

$$HOOC-CH_2-\underset{H}{\overset{H}{N}}: + \underset{Cl}{\overset{C_6H_5}{C}}=O + NaCH \xrightarrow{-Cl^-} HOOC\ CH_2\underset{H}{\overset{H}{N^+}}-\overset{C_6H_5}{C}=O$$

Glycine Benzoyl chloride

$$\downarrow -H^+$$

$HOOC\ CH_2NH\ COC_6H_5$

Benzoyl glycine
(Hippuric acid)

Materials

Glycine	5 g
Sodium hydroxide (10% solution)	50 mL
Benzoyl chloride	10.8 g (9 mL)
Carbon tetrachloride	40 mL

Procedure

Benzoyl chloride (9 mL, 0.078 mol) is added in two lots to a solution of glycine (5 g, 0.067 mole) in sodium hydroxide solution (10%, 50 mL). After each addition, the flask is stoppered and vigorously shaken. The mixture is cooled by addition of crushed ice and carefully acidified to corgo red with conc. HCl. The solid product in filtered, washed with water and air-dried. It is heated with carbon tetra chloride (2 × 20 mL) (to remove any contaminated benzoic acid) and crystallised from boiling water, m.p 187-188° yield 9 g (75.6%).

2.22 HYDANTION

It is obtained by heating glycerine with urea in presence at alkali, the formed hydantoic acid sod. salt on heating with conc. HCl yields hydrantion

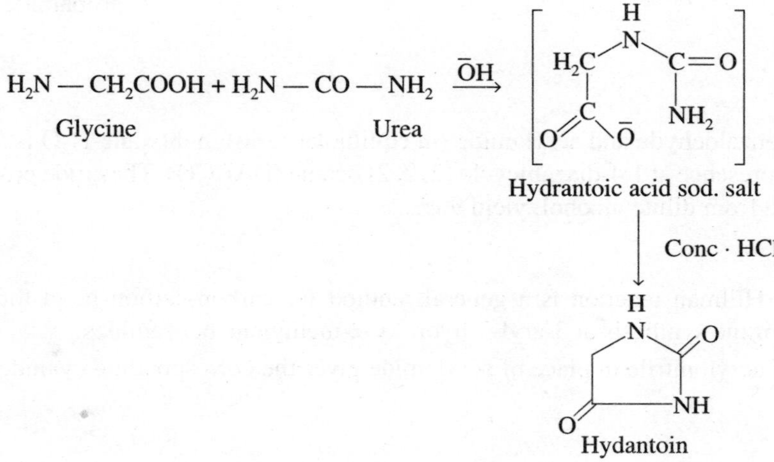

$$H_2N-CH_2COOH + H_2N-CO-NH_2 \xrightarrow{\bar{O}H}$$

Glycine Urea

Hydrantoic acid sod. salt

$$\downarrow \text{Conc} \cdot HCl$$

Hydantoin

Materials

Glycine	3.8 g
Urea	6 g
NaOH	2g
Conc. HCl	5 mL

Procedure

Aqueous sodium hydroxide solution (2 g in 3 mL H_2O) is added dropwise to a mixture of glycine (3.8 g, 0.05 mol) and urea (6 g, 0.1 mol) and the mixture heated at 110-115° (oil bath) for 1 hr. It is cooled, conc. HCl (5 mL) added and the mixture heated at 110-115° (oil bath) for 15 min. The reaction mixture is cooled, separated hydation filtered and crystallized from water. Yield 2.6 g (78.7%), m.p. 220-221°.

Notes

1. The reaction can also be conducted in microwave oven (2 min heating in both the steps), i.e., first by heating with aq. NaOH and then by heating with conc. HCl.

2.25 3-HYDROXY-3-PHENYL-2-METHYLENE PROPONAMIDE

It is obtained by the reaction of benzaldehyde with acrylamide in 1,4-dioxane-H_2O in the presence of DABCO (catalyst) at room temperature. The reaction is known as **Baylis-Hillman reaction** (J.H.Zhang C.M Wei, and C.J, Li., Tetrahedron Lett., 2002, **43**, 5731)

$$C_6H_5\,CHO + CH_2{=}CH{-}C(O)NH_2 \xrightarrow[\text{1,4-dioxane-}H_2O]{\text{DABCO, rt}} C_6H_5{-}CH(OH){-}C({=}CH_2){-}C(O)NH_2$$

Benzaldehyde   Acrylamide   80% yield 3-Hydroxy-3-phenyl-2-methylene propamide

Procedure

A mixture at benzaldehyde and acrylamide (in equimolar ratio) in dioxane-H_2O is stirred at room temperature in presence at 1,4-diazabicyclo [2, 2, 2] octane (DABCO). The crude product is filtered and crystallized from dilute alcohol, yield 90%.

Notes

1. Baylis-Hillman reaction is a general method for carbon-carbon bond formation and is used for the synthesis at 3-aryl-3-hydroxy-2-methylene benzamides
2. Use of acrylonitrile in place of acrylamide gives the corresponding cyanide

$$CH_2{=}CH\text{–}CN + Ph\,CHO \longrightarrow Ph(OH)C(=CH_2)CN$$

3. The reaction could be conducted in much shorted time by sonication
4. Baylis-Hillman reaction has also been conducted in polyethylene glycol (PEG-400) (as a rapid and recyclable reaction medium) with the conventional basic catalyst DABCO (20 mol %) giving very good yield (70-90%). In this case recyclability is achieved with no further addition at DABCO to the reaction medium for over 4 runs. This is the first reported recyclibility of DABCO (S. Chandrasekhar, Ch. Narsihmulu, B.Saritha and S.S.Sultana, Tetrahedron Lett., 2004, **45**, 5865). In this case, even aliphatic aldehydes (like 3-phenylpropanal, isobutyraldehyde, hexanal) could be used.
5. Baylis-Hillman reaction can also be conveniently performed by MW irradiation in high yields and reduction of reaction time (M.K. Kundu, S.B. Mukherjee, N. Bala, R. Padmakumar and S.V.Bhat, Synlett., 1994, 444.

$$\underset{\text{Benzaldehyde}}{C_6H_5CHO} + \underset{\text{Methyl crotonate}}{CH_3CH{=}CHCO_2Me} \xrightarrow[\text{MW, 2-3 min}]{\text{DABCO}} C_6H_5CH(OH)C(CO_2Me){=}CH\text{–}CH_3$$

2.24 IODOFORM

It is prepared from acetone by reaction with iodine in presence at sodium hydroxide. The reaction is known as **haloform reaction.**

$$CH_3COCH_3 \xrightarrow{^-OH} CH_3CO\bar{C}H_2 \xrightarrow{I_2} CH_3COCH_2I \xrightarrow{HO/I_2}$$

$$CH_3COCHI_2 \xrightarrow{^-HO/I_2} CH_3COCI_3 \xrightarrow{^-OH} H_3C\text{—}C(\ddot{O}:^-)(OH)\text{—}CI_3$$

$$\longrightarrow CH_3\text{—}\overset{O}{\overset{\|}{C}}\text{—}O^- + \underset{\text{iodoform}}{CHI_3}$$

Materials

Acetone	3mL
Sodium hydroxide solution	10%, 15 mL
Iodine solution	12.5 g iodine dissolved in a solution of 12.5 g KI in 50 mL water.

Procedure

Iodine solution (12.5 g iodine dissolved in a solution at 12.5 g. KI in 50 mL water) is added slowly with stirring to a solution of acetone (3 mL, 0.04 mol) in water (30 mL) and sodium hydroxide solution (10%, 15 mL). The mixture is heated at 60° (water bath) till the precipitate of iodoform settles down. It is filtered and crystallized from dilute methanol (1:1), M.P. 119°. yield 5 g (31.2%)

Notes

1. Haloform reaction is also known as **Liebsen iodoform reaction**. A.Lieben, Ann. (Suppl, 1870, **7**, 218; R.C.Fuson, B.A.Bull, Chem. Rev., 1934, **15**, 275.
2. The haloform reaction consist in the cleavage of methyl ketones (CH_3-CO-R), acetaldehyde, ethanol and secondary methyl carbinols (CH_3CHOH R) with halogens (mostly iodine) and a base to give haloform (iodofrom if iodine is used) and carboxylic acids.

$$RCOCH_3 + 3NaOI \rightarrow RCOCI_3 + 3\ NaOH$$

$$RCOCI_3 + NaOH \rightarrow CHI_3 + R\ COONa$$

3. Iodoform can also be prepared from ethyl alcohol (V.K. Ahluwalia and Renu Aggrwal, comrehensive practical organic chemistry: preparations and quantitative Analysis, Universities Press, 2004, Page 116)

$$C_2H_5OH + 4I_2 + 3K_2CO_3 \rightarrow CHI_3 + HCOO^-K^+ + 5KI + 3CO_2 + 2H_2O$$

2.25 INODOLE

It is obtained by the decarboxylation of indole-2-carboxylic acid in quantitative yield in 20 min by heating in specially designed MW oven at 255°. This region at 255° is known as near critical region (J. An, T. Bangelli, T. Cablewski, C.R.Straus and R.W.Trainor, J. Org.Chem., 1997, **62**, 2502).

Indole 2-carboxylic acid (–CO_2H, N–H) $\xrightarrow[\text{MW, 255°, 20 min}]{H_2O}$ Indole (93%) (N–H)

Indole 2-carboxylic acid — Indole (93%)

Even 2-carbethoxy iodole on similar hydrolysis in 0.2 M aq. NaOH also gave indole, in this case the ester group is first deesterified, which undergoes decarboxylation to give indole in 93% yield.

2.26 3-(p-METHOXYPHENYL)-2H-1,4-BENZOXAZINE

It is prepared by the reaction of o-amino phenol with p-methoxy phenacyl bromide in dichoromethane in presence at tetrabutylammonium hydrogen sulphate.

OH, NH_2 + $BrCH_2$ — CO — — OCH_3 $\xrightarrow[\text{aq} \cdot K_2CO_3,\ CH_2Cl_2]{n\text{-}Bu_4\overset{+}{N}HSO_4^-}$

o-Aminophenol   p-Methoxy phenacyl bromide

⟶ O, N, OCH_3 60%

3-(p-Methoxyphenyl)-2H-1,4-Benzoxazine

Procedure[1]

Stir a mixture of o-aminophenol (0.01 mol) in dichloromethane (50 mL), aqueous potassium carbonate (20% 40 mL) and tetrabutlammonium hydrogen sulphate (0.0005 mol). To the stirred mixture add a solution of p-methoxyphenacyl bromide (0.01 mol) in dichloro methane (20 mL) during 15-20 min. Stirr the mixture for 1 hr, separate the organic layer, wash with water (2 × 100 mL), dry (Na_2SO_4) and evaporate. Crystallise the residual product from alcohol, m.p., 131°, yield 61%

Notes

1. G. Sabitha and A.V.Subba Rao, Synth. Commun., 1987, **17**, 341.
2. The PTC, Tetrabutyl ammonium hydrogen sulphate is prepared described in see Section 5.8 Note 3

2.27 3-METHYLCYCLOPENT-2-ENONE

It is obtained by the reaction of hexane –2, 5 dione by an **intermolecular aldol condensation.**

CH_3, C=O, H_2C, CH_3, H_2C—C=O $\xrightarrow{\bar{O}H}$ CH_3, C=O, H_2C, $\ddot{C}H_2^{\ominus}$, H_2C—C=O ⟶ [H_3C, OH, C, H_2C, CH_2—C=O] $\xrightarrow{-H_2O}$ CH_3, O

Hexane-2,5-dione   3-Methylcyclopent-2-enone

Materials

Hexane – 2, 5-dione	2.85 g
Sodium hydroxide	0.25 g

Procedure

To a boiling solution of sodium hydroxide (0.25 g) in water (25 mL) is added dropwise hexane-2,5-dione (2.85 g, 0.025 mol). After the addition, the reaction mixtures is kept boiling for 15 minutes, cooled rapidly (ice-salt mixt.), saturated with sodium chloride, extracted with either (3 × 20 mL). The ether extract is washed with water (2 × 3 mL), dried (anhyd. Na_2SO_4) and ether removed by distillation. 3-Methylcydopent-2-enone is obtained as an oily product, b.p. 74-76°/10 mm, yield 0.95 g (40%)

Notes

1. The required hexane-2, 5-dione is obtained form ethyl acetoacetate as given below:

$$CH_3CO \cdot CH_2 \cdot CO_2\ C_2H_5 \xrightarrow{2Na} 2\left[CH_3 \cdot CO \cdot CH \cdot CO_2C_2H_5\right]^{\ominus} \overset{\oplus}{Na}$$

$$\xrightarrow{I_2} \begin{array}{l} CH_3 \cdot CO \cdot CH \cdot CO_2C_2H_5 \\ \qquad\qquad\quad | \\ CH_3 \cdot CO \cdot CH \cdot CO_2C_2H_5 \end{array} \xrightarrow[\text{decarboxylation}]{\text{hydrolysis}} CH_3 \cdot COCH_2 \cdot CH_2 \cdot CO \cdot CH_3$$

2. For the preparation of 3-methyleyclopent-2-enone, it is necessary to follow to the exact reaction conditions in order to avoid the formation of tarry byproducts (in case excess of concentrated alkali is used).

2.28 2-(2′-METHLINDOL-3YL)-1,4-BENZOQUINONE

It is obtained by the reaction 2-methyindole with 1,4-benzoquinone in water.

2-Methylindole + p-Benzoquinone $\xrightarrow[\text{r.t. 10 hr.}]{H_2O}$ 2-(2′-Methylindol-3yl)-1,4-benzoquinone 82%

Procedure

A mixture of 2-methylindole and p-benzoquinone (molar ratio 1:1) is stirred in water at room temperature for 10 hr. The product formed is filtered and crystallized from dilute alcohol yield 82%.

Notes

1. H.B. Zhong, L.Liu, Y.-J. Chen, D.Wang and C.-J Li, Eur J.Org, Chem, 2006, 869.
2. The reaction could be conducted is much shorter time using sonication.
3. The yields were very poor if the reaction is conducted in organic solvents like CH_2Cl_2, ether or THF.
4. In case the reactants are taken in the molar ratio 2:1, the product obtained is bis (indolyl)-1,4-benzoquinane.

O Cl N H CH_3 + Cl O $\xrightarrow[48\ h]{H_2O}$ O Me H N Cl Cl N H CH_3 O 73%

2.29 2-METHYL-2-(3-OXOBUTYL)-1,3-CYCLOPENTANEDIONE

It is prepared by the reaction of 2-methylcyclopentane-1,3-dione with methyl vinyl ketone in water in 87% yield. The reaction is a **Michael addition**[1], which takes place under acidic conditions due to enolic nature of the dione

H_3C O O $\longrightarrow$ H_3C O HO + $CH_2=CH-\overset{\overset{\displaystyle O}{\|}}{C}-CH_3$ $\xrightarrow{H_2O}$ H_3C O O O

2-Methylcyclopentane-1,3-dione — Methyl vinyl ketone — 2-Methyl-2-(3-oxybutyl)-1,3-cyclopentanedione

Materials

2-Methylcyclopentane-1,3-dione	6.5 g
Methyl vinyl ketone	9.6 mL
Water	14 mL

Procedure

A stirred suspension of 2-methylcyclopentane-1,3-dione (6.5 g) in water (14 mL) is treated with methyl vinyl ketone (9.6 mL). The stirring is continued at 20° for 5 days. The reaction mixture is extracted with hot benzene, benzene extract dried ($Na_2\ SO_4$) and distilled to give 2-methyl-2-(3-oxobutyl)-1, 3-cyclopentanedione as an oily product[2] (10 g).

Note

1. Z.G. Hojas and D.R. Parresh, J.Org. Chem., 1974, **39**, 1612
2. The product is characterised on the basis of its 1 R spectra (1770 and 1725 cm^{-1}) and NMR spectral data [δ 1.12 (s, 3H, 2-CH_3), 2.22 (s, 2H, CH_2 CO^-) and 2.82 (m, 4H, $COCH_2$ CH_2 CO^-)].
3. 2-Methyl-2(3-oxobuty)-1, 3-cyclopentanedione) was earlier reported to be prepared by the reaction of the dione and methyl viny ketone as a solid, m.p. 117–118° (C.B.C. Boycl and J.S. Whitehurst, J. Chem. Soc., 1959, 2022). However, the compound was not the required product (Z.G. Hajos and D.R. Parish, J.Org Chem., 1974, 1612).
4. In case, the above reaction is carried out in presence of a basic catalyst, the adduct cyclises to give fused ring system.

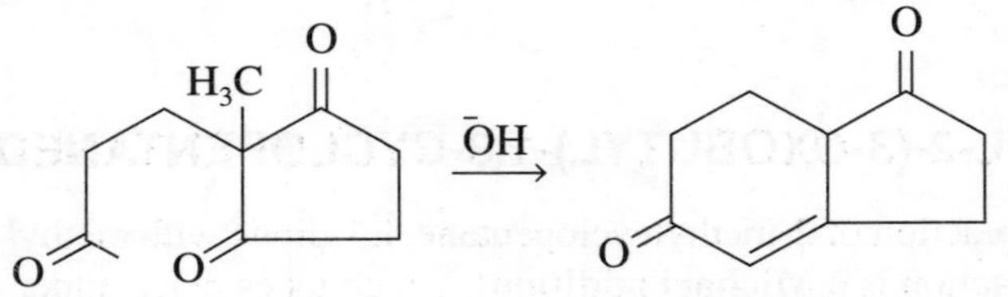

H_3C O O O $\xrightarrow{\bar{O}H}$ O O

2.30 β-NAPHTHYL ACETATE

It is obtained by the acetylation of β-naphthol with acetic anhydride in presence of aqueous NaOH.

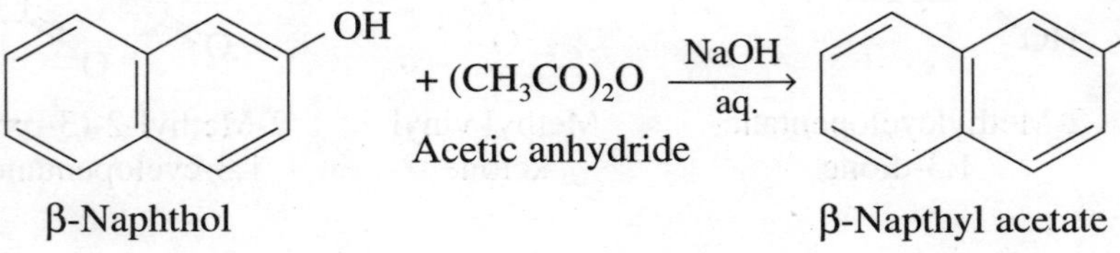

OH + $(CH_3CO)_2O$ $\xrightarrow[\text{aq.}]{\text{NaOH}}$ $OCOCH_3$

Acetic anhydride

β-Naphthol β-Napthyl acetate

Procedure

To a cooled solution (0-5°) of β-naphthol (4 g, 0.028 mol) in sodium hydroxide solution (10%, 20 mL) is added acetic anhydride (4.6 mL). The reaction mixture is shaken vigorously for 20-25 min. The separated β-naphthyl acetate is filtered, washed with water and crystallised from dilute alcohol, m.p. 71°. yield 5 g (96.8%)

2.31 β-NAPHTHYL METHYL ETHER (NEROLIN)

It is obtained by the methylation of β-naphthol by dimethyl sulpate in presence of sodium hydroxide

$$\text{β-Naphthol (OH)} \xrightarrow[\text{alkali}]{(CH_3)_2\,SO_4} \text{β Napthyl methyl ether } (OCH_3)$$

β-Naphthol β Napthyl methyl ether

Materials

β-Naphthol	9 g
Sodium hydroxide	2.65 g
Dimethyl sulphate	5.87 mL

Procedure

Dimethyl sulphate (5.87 mL, 0.062 mol) is added dropwise to a stirred and cooled solution (10-15°) of β-naphthol (9 g, 0.125 mol) in sodium hydroxide solution (2.65 g, 0.10 mol in 38 mL water). After the addition is over, the mixture is warmed at 70-80° for 1 hr and then cooled. The separated β-naphyl methyl ether in filtered, washed with sodium hydroxide solution (10%), water, dired and crystallised from alcohol, m.p. 72°. yield 8.25 g (84%)

Notes

1. Using the same procedure, resorcinol, 1,2,4-triacetoxy benzene and vanillin could be methylated to give resorcinol demethyl etter (b.p. 216-217°), 1, 2, 4-trimethoxybenzene (b.p. 247°) and veratraldehyde (m.p. 42-43°)

2.32 NAPHTHALDEHYDE

It is obtained by the reaction of 1-(chloromethyl) naphthalene with hexamine in dilute acetic acid.

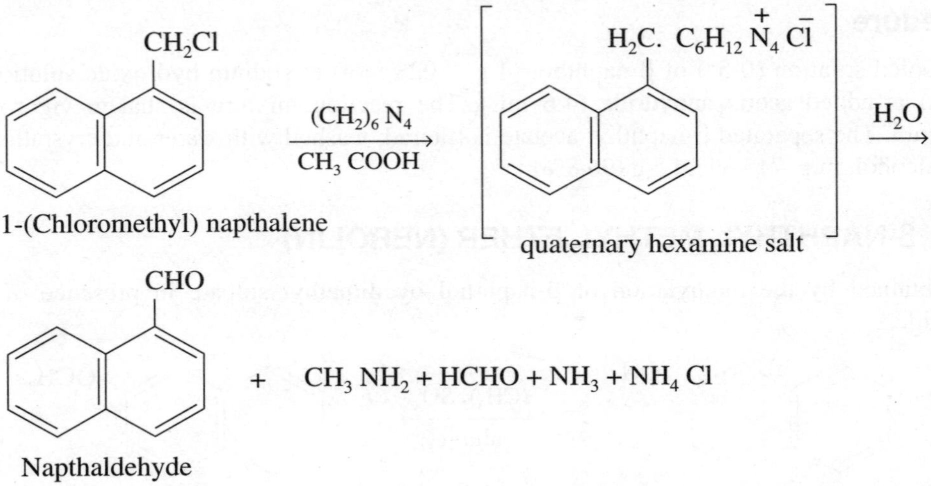

$$\text{1-(Chloromethyl) napthalene } (C_{10}H_7CH_2Cl) \xrightarrow[CH_3\,COOH]{(CH_2)_6\,N_4} \left[H_2C.\ C_6H_{12}\overset{+}{N}_4\,\overset{-}{Cl} \right] \text{ (quaternary hexamine salt)} \xrightarrow{H_2O}$$

$$\text{Napthaldehyde } (C_{10}H_7CHO) + CH_3\,NH_2 + HCHO + NH_3 + NH_4\,Cl$$

1-(Chloromethyl) napthalene

quaternary hexamine salt

Napthaldehyde

The reaction is known as **Sommelet Reaction** (M. Sommelet, Compt. Rend., 1913, **157**, 852; M. Sommelet, Bull. Soc. Chim. France, 1918, **23**, 95).

The quaternary hexamine salt formed form 1-(chloromethyl) naphthalene and hexamine (as shown above) undergoes hydrolysis with hot aqueous acid. The mechanism involved is given below:

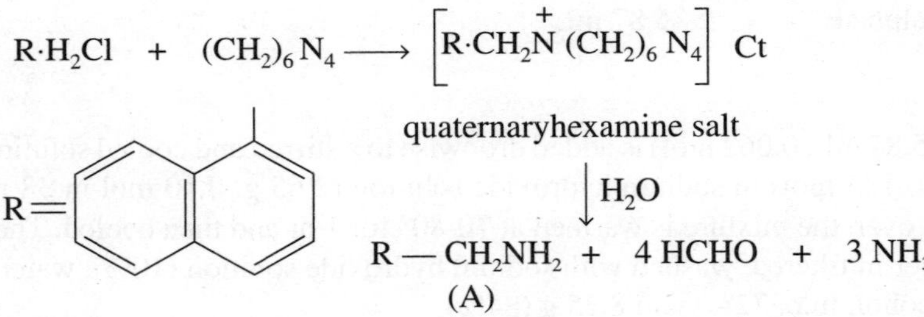

$$R{\cdot}H_2Cl \; + \; (CH_2)_6\,N_4 \longrightarrow \left[R{\cdot}CH_2\overset{+}{N}\,(CH_2)_6\,N_4 \right] Ct$$

quaternaryhexamine salt

R= (1-naphthyl)

$$\downarrow H_2O$$

$$\underset{(A)}{R - CH_2NH_2} \; + \; 4\,HCHO \; + \; 3\,NH_3$$

$$HCHO \; + \; NH_3 \longrightarrow H_2C{<}^{NH_2}_{OH} \xrightarrow[+H^+]{-H_2O} H_2C = \overset{+}{N}H_3$$

Protanated aldimine deriv.

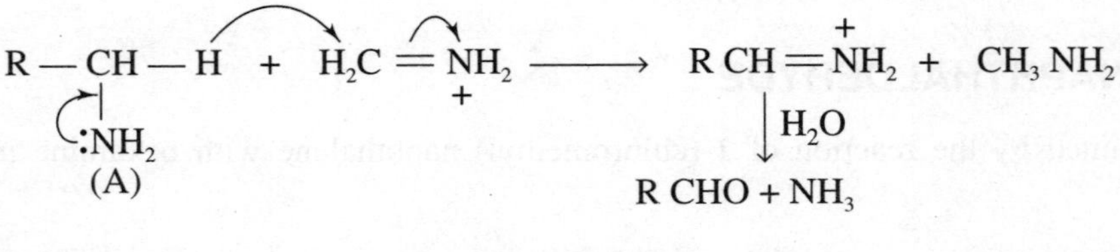

$$\underset{\underset{(A)}{:NH_2}}{R - CH - H} \; + \; H_2C = \overset{+}{N}H_2 \longrightarrow R\,CH = \overset{+}{N}H_2 \; + \; CH_3\,NH_2$$

$$\downarrow H_2O$$

$$R\,CHO + NH_3$$

Materials

1-(Chloromethyl) naphthalene	5.3 g
Hexamine (hexamethylenetetramine)	8.4 g
Acetic acid (aqueous) (50%)	25 mL

Procedure

A mixture of 1-(chloromethyl) naphthalene (5.3 g, 0.03 mol), hexamine (8.4 g, 0.06 mol) and aqueous acetic acid (50%, 25 mL) is refluxed for 2 hr. Conc. HCl (10 mL) is added and mixture refluxed for 15 min more. The cooled mixture is extracted with ether (2 × 10 mL), ether extract washed with water (2 × 5 mL), sodium carbonate solution (10% 5 mL) and finally with water. The ethereal solution is dried (anhydrous $MgSO_4$) and distilled. 1-Naphthaldehyde, b.p. 160-162°/10 mm is obtained in 81% yield (3.8 g).

Notes

1. This is a general method for preparing aromatic aldehydes from the corresponding methyl compound by heating the chloromethyl or bromomethyl compound with hexamine and hydrolysis of the formed quaternary hexamine salt with hot aqueous acetic acid.

2.33 2-NAPHTHOIC ACID

It is obtained by the **haloform reaction** involving oxidation 2-acetylnaphthalene with sodium hydrochlorite. For mechanism sec Section 2.24 (preparation of iodoform)

$COCH_3$ —NaOCl / HCl→ CO_2H

2-Acetylnapthalene — 2-napthoic acid

Materials

2-Acetylnaphthalene	4.25 g
Sodium hypochlorite solution containing 10-15% available chlorine (commercially available)	30 mL
Sodium metabisulphite	2.5 g

Procedure

To a solution of sodium hypochlorite (30 mL) in water (45 mL) at 50° is added slowly 2-acetyl naphthalene (4.25 g, 0.025 mol) by frequent stirring and cooling (ice-bath) during 30 min. The mixture is stirred for 30 min. more; excess hypochlorite destroyed by adding a solution of sodium metabisulphite (2.5 g) in water (10 mL). The cooled reaction mixture is acidified with conc. HCl (10 mL), separated product filtered, dried (100°) and crystallised from ractified spirit, m.p. 184-85° yield 3.75 g (87%)

Notes

1. 2-Acetyl naphthalene is commercially available

2.34 PHENYL ACTATE

It is prepared by the reaction of phenol with acetic anhydride in alkaline medium

$$C_6H_5OH + (CH_3CO)_2O \xrightarrow[\text{stirred}]{\text{NaOH solution}} C_6H_5OCOCH_3 + CH_3COONa$$

Phenol + Acetic anhydride → Phenyl acetate (95%)

Materials

Phenol	9.4 g
Sodium hydroxide solution 10%,	64 mL
Acetic anhydride	12.0 mL

Procedure

To a vigorously stirred solution of phenol (9.4 g, 0.1 mol), sodium hydroxide solution (64 mL, 10%) and crushed ice (75g) is added slowly during 5 min, acetic anhydride (12.0 mL, 0.127 mol). The reaction is complete in about 10 min. To the reaction mixture is added carbon tetrachloride (35-40 mL) (to separate the resulting emulsion). The lower layer of phenyl acetate is separated (separatory funnel), washed with dilute sodium carbonate solution, dried ($CaCl_2$) and distilled. Yield 13 g (95%), b.p. 194-197°.

Notes

1. Phenyl acetate can also be prepared by **Baeyer-Villiger oxidation** of acetophenone with peracids (A.V. Balyer and V. Villiger, Ber., 1899, **32**, 3625). The mechanism of the reaction is given below.

$$C_6H_5-\overset{O}{\overset{\|}{C}}-CH_3 \underset{}{\overset{-H^+}{\rightleftharpoons}} \left[C_6H_5-\overset{\overset{+}{O}H}{\overset{\|}{C}}-CH_3 \longleftrightarrow C_6H_5-\overset{OH}{\overset{|}{\underset{+}{C}}}-CH_3\right] \xrightarrow[-H^+]{C_6H_5CO_3H} C_6H_5-\overset{\ddot{O}H}{\overset{|}{C}}(-CH_3)-O-O-C(=O)-C_6H_5$$

$$\xrightarrow{-C_6H_5COO^-} \left[C_6H_5-\overset{\overset{+}{O}-H}{\overset{\|}{C}}-OCH_3\right] \xrightarrow{-H^+} C_6H_5-\overset{O}{\overset{\|}{C}}-OCH_3$$

Phinyl acetate

2. Baeyer-Villiger oxidation can also be carried out in solid state. The procedure consist in keeping an intimate mixture of powered ketone and 2 equivalent of m-chloroperbenzonic acid at room temperature for 24 hr. (K. Tank and F. Toda, Chem. Rev., 2000, **100**, 1028)

$$\underset{\text{Benzyl phenyl ketone}}{PhCOCH_2Ph} \xrightarrow[\text{MCPBA solid state}]{\text{RT, 24 hr}} \underset{\text{97\% Benzylbenzoate}}{PhCO\,OCH_2Ph}$$

3. Baeyer-villiger oxidation is also be carried out using enzymes. Thus, phenylacetone on treatment with the enzyme cyclohexanone oxygenase gives benzyl acetate (B.P. Branchaud and C.T. Walsch, J. Am. Chem. Soc., 1985, **107**, 2153)

$$\underset{\text{Phenyl acetone}}{C_6H_5CH_2COCH_3} \xrightarrow[\text{Enz-FAD, NAD PH, } H^+]{\text{Cyclohexanone oxyganase, } O_2} \underset{\text{Benzyl acetate}}{C_6H_5CH_2OCOCH_3}$$

2.35 PHENYLACETIC ACID

It is obtained by the hydrolysis of benzylcyanide with dilute sulphuric acid in acetic acid.

$$\underset{\text{Benzyl cyanidle}}{C_6H_5-CH_2-C\equiv N} + H-O-H \xrightarrow{-OH^{\ominus}} C_6H_5\,CH_2C(OH){=}NH \longrightarrow C_6H_5\,CH_2-C(=O)-NH_2$$

$$C_6H_5\,CH_2C(=O)-NH_2 + H-O-H \xrightarrow[-NH_3]{-\overset{\ominus}{O}H} C_6H_5\,CH_2-C(=O)-OH$$

Materials

Benzyl cyanide	5.4 mL
Conc. H_2SO_4	5 mL
Glacial acetic acid	5 mL.

Procedure

A mixture of benzyl cyanide (5.4 mL, 0.047 mol), water (5 mL), glacial acetic acid (5 mL) and conc. H_2SO_4 (5 mL) is refluxed for 45 min in a R.B flask (50 mL capacity). The reaction mixture is poured into crushed ice (70-75 g). The separated phenylacetic acid is filtered, washed with cold water and crystallised form hot water m.p. 77°, yield 1.2 g (37.7%).

Notes

1. The required benzyl cyanide is prepared from benzyl chloride by reaction with KCN in presence of PTC (see Section 5.2a).
2. Phenyl acetic acid can also be obtained by the reaction of grignard reagent, $C_6\ H_5CH_2$ MgCl with carbon dioxide.

2.36 2-PHENYL p-BENZOQUINONE

It is obtained by the coupling of 1,4-benzoquinone with benzene in water (H-B. Zhang, L. Liu, Y.–J. Chen, D. Wang and C.–J. Li, Adv. Synth. Catal. 2006, 348, 229).

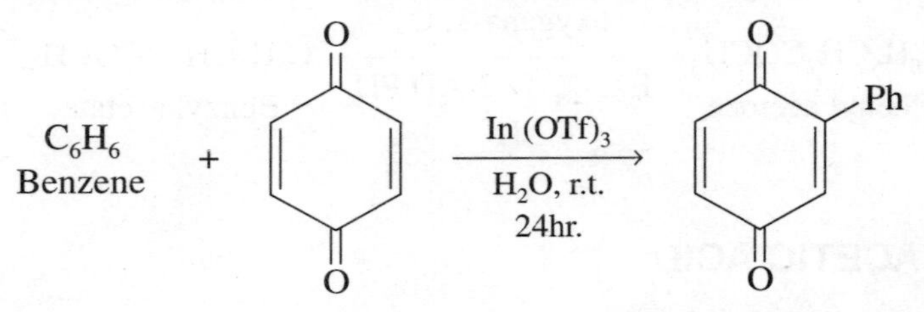

2-Phenyl p-benquinone
75%

Procedure

A mixture of benzene and p-benzoquenone (in the ration 1:2) in water is stirred for 24 hr. in presence of the catalyst, In $(OTF)_3$ (5 mol%). The coupled product is filtered and crystallised from water. yield 75%.

Notes

1. This is a general procedure for arylation of 1,4-quinones in water
2. Usingtheprocedure,anumberofaromaticcompounds(like2-methoxy-N,N-dimethylaniline, N, N-dimethylaniline 1-N, N-dimethylnaphthalene and N-methylaniline) could be coupled to give 2-subsituted 1,4-benzoquinones in 80-90% yield. Even 1,4-naphthoquinones could be condensed with aromatic compounds to give the coupled products.
3. The reaction could be completed in much sorter time using sonication.

2.37 1-PHENYL-2, 2-DICHLOROCYCLOPROPANE

It is prepared by the addition of dichlorocarbene (genenerated in situ from chloroform, aqueous sodium hydroxide and a PTC, benzyltriethlyammonium chloride) to styrene.

$$\text{Styrene} \xrightarrow[\left[C_6H_5CH_2\overset{+}{N}(C_2H_5)_3\right]Cl^-,\ 40°,\ 4h]{CHCl_3,\ \text{aq. NaOH}} \text{1-Phenyl-2,2-dichlocyclopropane (80\%)}$$

Procedure[1]

Stir a mixture of styrene (0.1 mol), chloroform (0.1 mol), aqueous sodium hydroxide (50% 20 mL) and benzyltriethylammonium chloride (0.4 g) for 4 hr. at 40°. Dilute the reaction mixture with water and extract with ether. Dry ether extract over anhydrous magnesium sulphate and distil to give 1-phenyl-2,2-dichorocyclopropane in 80% yield.

Notes

1. M. Makosza and M. Waworzniewicz, Tetrahedron Lett., 1969, 4659.
2. The required PTC, viz. benzyltriethylammonium chloride was prepared as given in Section 5.1.3 (Note 2)
3. 1-Phenyl-2, 2-dichlorocyclopropane can also be prepared by reaction of styrene with chloroform in presence of solid (powdered) sodium hydroxide using sonication (S.L. Regen and A.K. Singh, J. Org. Chem. 1982, **47**, 1587)

$$\text{Styrene} \xrightarrow[\text{))))},\ 4\ \text{hrs},\ 40°]{NaOH,\ CHCl_3} \text{1-Phenyl-2,2-dichlorocyclopropane}$$

Much better yields (96%) are obtained by both mechanical stirring and ultrosonic irradiation.

2.38 β-PHENYLPROPIONIC ACID

It is obtained by the reduction of cinnamic acid with dimide generated in situ by the reaction of hydrazine hydrate with hydrogen peroxide in presence of cupric sulphate

$$\underset{\text{Cinnamic acid}}{C_6H_5CH = CH\ COOH} \xrightarrow{HN = NH} \underset{\beta\text{-Phenyl propionic acid}}{C_6H_5CH_2CH_2\ COOH + N_2}$$

Materials

Cinnamic acid	4 g
Hydrozine hydrate	8 g
Cupric sulphate	50 mg
Hydrogen peroxide (30%)	10 g (9 mL).

Procedure

To a cooled solution (ice-bath) of cinnamic acid (4 g, 0.027 mol) in hydrazine hydrate (8 g), water (50 mL) and cupric sulphate (50 mg) is slowly added during 20-25 min, hydrogen peroxide (30%, 10 g, 9 mL). The temperature is kept below 30°. The resulting mixture is kept at room temperature for 10 min. and then cooled. To the cooled mixture (ice-bath) is added hydrochloric acid (1:1) with stirring. The separated β-phenylpropionic acid is filtered and crystallised from ether m.p. 48-49° yield 1.6g (39.5%)

2.39 PINACOLONE

It is obtained by the rearrangement of pinacol by heating in MV oven at 270° in water (J.M. Kremsner and C.O. Kappe, Eur. J. Org. Chem., 2005, 3672)

$$\underset{\text{Pinacol}}{\mathrm{Me_2C(OH){-}C(OH)Me_2}} \xrightarrow[\text{MW, 270°, 30 min}]{H_2O} \underset{\text{Pinacolone (76\%)}}{\mathrm{Me_3C{-}CO{-}Me}}$$

The transformation is called **Pinacol-Pinacolone rearrangement**.

$$\underset{\text{Pinacol}}{\mathrm{Me_2C(OH){-}C(OH)Me_2}} \xrightarrow{H^+} \mathrm{Me_2C(OH){-}C(\overset{+}{O}H_2)Me_2} \xrightarrow{-H_2O} \mathrm{Me_2C(OH){-}\overset{+}{C}Me_2}$$

$$\xrightarrow{\text{1,2-shift}} \mathrm{Me{-}\overset{+}{C}(OH){-}CMe_3} \longrightarrow \mathrm{Me{-}C({=}\overset{+}{O}H){-}CMe_3} \xrightarrow{-H^+} \underset{\text{Pinacolone}}{\mathrm{M{-}C({=}O){-}CMe_3}}$$

Procedure

Pinacol is heated in water at 270° in a microwave oven for 30 min. Pinacolone is obtained in 76% yield. The product was isolated as 2, 4-dinitrophenyl hydrazone by treatment with 2, 4-dinitrophenyl hydrazine. Alternatively the product can be isolated by steam distillation b.p. 106-107°.

Notes

1. The required pinacol is obtained by the reduction of acetone with amalgamated magnesium in 67.4% yield. M.P. 45-46° (E.W. Adams, Org. Synth. Coll. Vol. I, 459.
2. The reaction can also be conducted by heating pinacol in H_2O in a sealed tube at 270° for 30 min.
3. The pinacol-pinacolone transformation normally proceeds by heating pinacol with 25% H_2SO_4 for 3 hrs. (Hill, Flosdorf, Org. Synth. Coll. Vol-I, 1941, P 462)
4. In the above transformation water at about 270° (is near critical water, (NCW) acts as an acid catalyst itself).
5. Pinacol-pinacolone rearrangement has also been carried out in a solid state using microwave irradiation (E. Gutierrez, A. Loupy, G. Bram and E. Ruiz-Hitzkey, Tetrahedron Lett., 1989, **30**, 945

OH
HO
Al^{3+}–montmorillonite
MW, 15 min
O

Pinaciol

Pinaeolone
98-99%

6. Pinacol-pinacolone rearrangement can also be performed by heating pinacol in ionic liquid at about 150° for 2 hr. (A. Cole, J.L Jensen, I. Ntai, K.L.T. Tran, KJ. Weaver, D.C. Forbes and J.H. Davis, J. Am. Chem. Soc., 2002, 124, 5962.

$$\text{Ionic liquid} \rightarrow \geq p^{+} (CH_2)_3\ SO_3H\ p\text{-}CH_3\ C_6H_4\ SO_3^{-}$$

7. See also preparation of benzopinacolone (Section 12.6).

2.40 SALICYLIC ACID

The alkaline hydrolysis of esters is called **saponification**. For example, methyl salicylate on saponification with sodium hydroxide solution gives sodium salicylate, which on acidification gives salicylic acid.

$$R-\overset{\overset{O}{\|}}{C}-OCH_3 \longrightarrow R-\underset{OH}{\overset{O^-}{|}}{C}-OCH_3 \longrightarrow RCOOH + CH_3O^-$$

$$\uparrow\ ^-OH \qquad\qquad \downarrow H^+ \text{ exchange}$$

$$RCOOH + CH_3OH \xleftarrow{H^+} RCOO^- + CH_3OH$$

$R=$ 2-methoxyphenyl (OCH_3 substituted benzene ring)

Materials

Methyl salicylate	2 mL
Sodium hydroxide 10%	15 mL

Procedure

A mixture of methyl salicylate (2 mL) and sodium hydroxide solution (15 mL, 10%) is refluxed on a sand bath (90-100°) using a reflux condenser for about 30 min till the ester layer disappears. The solution is cooled, acidified (dil. HCl) and cooled (ice-bath). The separated salicylic acid is filtered and crystallised from hot water yield 1.2 g (90%) m.p. 158-159.

Notes

1. Use of a crown ether, viz. [18] crown-6 in small amount for saponification gives quantitative yield. The special feature of using crown ether, is that even the sterically hindered esters, which are difficult to saponify with alkali can be saponified conveniently by using [18] crown 6. (C.J. Pedersen and K.K. Friensdorff, Angew Chem. Int. Engl., 1972, **11**, 16; C.J. Pedersen, J. Am. Chem. Soc, 1967, **89** 2485, 7017; 1970, **92**, 386, 391).
2. Saponification of esters can also be brought about by using microwave under solid-liquid PTC conditions without solvent. The reaction can be completed in few minutes. For example,

$$C_6H_5-\overset{\overset{O}{\|}}{C}-OCH_3 \xrightarrow[\text{7 min microwave, aliquat 336}]{KOH} C_6H_5-\overset{\overset{O}{\|}}{C}-OK \quad 94\%$$

(A. Loupy, P. Pigeon, M. Ramdani, P. Jaequault, Synthetic Common., 1994, **24**(2), 159).

3. Saponification can also be conducted under milder conditions when sonication in used (S. Moon, L. Duclin, J.V. Croney, Tetrahedron Lett., 1979, 3971). In this case rate increase in attributed the emulsifying effect. A typical example is given below.

$$\text{Methyl 2,4-dimethyl benzoate (COOCH}_3\text{, CH}_3\text{, CH}_3) \xrightarrow{^-OH/H_2O,))),\ 60\ min} \text{2,4 Dimethyl benzoic acid (COOH, CH}_3\text{, CH}_3)\ (94\%)$$

Methyl 2,4-dimethyl benzoate — 2,4 Dimethyl benzoic acid (94%)

2.41 STILBENE (TRANS)

Stilbene (a mixture of cis and trans forms) was earlier prepared (G. Markl and A. Merz, Synthesis, 1973, 295) by the reaction of benzaldehyde with benzyltriphenyl phosphonium chloride. The reaction is known as **Witting Reaction**[1].

$$(C_6H_5)_3\ddot{P} + C_6H_5CH_2-Cl \xrightarrow{-Cl^-} (C_6H_5)_3P\ CH_2C_6H_5 \xrightarrow[-BH]{:B} (C_6H_5)_3\overset{+}{P}-\overset{-}{C}H\ C_6H_5$$

Triphenyl phosphine — Benzyl chloride — Benzyl Triphenyl phosphosphonium salt

$$(C_6H_5)_3\overset{+}{P}-\overset{-}{C}H\ C_6H_5 + \underset{H}{\overset{C_6H_5}{>}}C=O \xrightarrow{\Delta} (C_6H_5)_3\overset{+}{P}-\overset{C_6H_5}{\overset{|}{C}H}-\underset{O^-}{\overset{H}{C}}-C_6H_5 \longrightarrow$$

Benzaldehyde

$$\longrightarrow \underset{(C_6H_5)_5\,P\,-\,O}{HC(C_6H_5)-C(H)-C_6H_5} \longrightarrow \underset{H}{\overset{C_6H_5}{>}}C=C\underset{C_6H_5}{\overset{H}{<}} + (C_6H_5)_3\,P=O$$

Stilbene (cis + trans) — Triphenyl phosphine oxide

It has now been possible to carry out the above Witting reaction in presence of aqueous sodium hydroxide[2]

Materials

Benzaldehyde	4.24 g
Benzyltriphenyl phosphorium chloride	15.72 g
Methylene chloride	20 mL
Sodium hydroxide solution	50%, 20 mL

Procedure

A mixture of benzaldehyde (4.24 g), benzyltriphenyl phosphorium chloride (15.72 g) and methylene chloride (20 mL) is stirred and treated with aqueous sodium hydroxide solution (20 mL 50%) (added slowly using a separatory funnel). The temperature rises to 50°. The mixture is kept at 50° for 30 min with vigorous stirring. The organic layer is separated, washed with water (25 mL), sodium bicarbonate solution (40 mL) and finally with water. The extract is dried (Na_2SO_4), filtered and evaporated. The trans stilbene is crystallised from ethanol yield 2 g. M.P. 122-123°.

The mother liquor (left after crystallisation of trans stilbene is treated with petroleum ether (b.p. 40-60°, 40 mL). The precipitated triphenyl phospine oxide (10g) (m.p. 146-47°) is filtered and the filtrate is evaporated (vacuo, room temperature). The residual oily product consisted of cis-stilbene (3g). On cooling (–5°) it solidified, b.p. 135°/10 mm.

Notes

1. C. Wittig and U. Schollpoff, Bors, 1954, **87**, 1318; C. Wittig and N. Haag, Ber, 1955, **88**, 1654.
2. Johy C. Warner, Pault Anastas and Jean-Pierre Anselme, J. Chem. Education, 1985, **62**, 346.
3. The Benzyl triphenyl phosphonium chloride required is prepared by refluxing a mixture of benzyl chloride (4.4 ml), triphenyl phosphine (14.3 g) and xylene (70 ml). The separated product is filtered, m.p. 310-11°. It is washed with xylene and dried in vacuum dessicator.
4. If possible take the nmr spectra of cis and trans stilbene and characterise both compounds.
5. The above procedure of Witting condensation can also be performed with trans-cinnamaldehyde, p-metyl--, p-methoxy-, p-chloro-, m- and p-nitrobenzaldehydes and 9-anthracenealdehyde. The product is obtained in 65-85% yield (G. Markl and A. Merz, synthesis, 1973, 295).
6. Substituted stilbenes can also be obtained by **Heck reaction** in water using MW. heating (R.K. Arvela, and N.E. Lead beater, J. Org. Chem. 2005, **70**, 1786). The method consists in heating aryl bromides with styrene in presence of catalytic amount of Pd. (concentration as low as 500 ppb), Na_2CO_3 tetrabutylammonium bromide (TBAB), H_2O, M.W.

Br, R + Pd, Na_2CO_3, TBAB / H_2O, MW-170°, 10 min → R

Arylbromide
R = Me, OMe, COME

Stilbene

Trans stilbene
60-70%

7. The Wittig reaction has also been reported to proceed in solid state in presence of a PTC (F. Toda and K. Akai, J. org. Chem., 1990, **55**, 3446)
8. Ionic liquids have also been used to perform Witting reaction (V.L. Boulaire and R. Gree, Chem. Commun., 2000, 2195. The advantage of using ionic liquid is easy separation of alkenes from Ph_3PO and also recyclising of the solvent.

2.42 2,3,4-TRIMETHOXYPHENOL

It is obtained by the oxidation of 2,3,4-trimethoxybenzaldehyde with peracetic acid followed by hydrolysing the formed ester with alkali. This oxidation is known as **Baeyer-Villiger oxidation**. The mechanism of oxidation is given below:

$$\mathrm{RHC{=}O} \xrightarrow{H^+} \mathrm{RHC{=}\overset{\oplus}{O}H} \xrightarrow{CH_3COOO^-} \mathrm{RHC(O{-}O{-}OCH_3)(O{-}H)} \longrightarrow$$

$$\longrightarrow \mathrm{H{-}C(=O)OR} + CH_3COO^- + H^+$$

$$\mathrm{H{-}C(=O)OR} \xrightarrow[OH,\ \Delta]{\text{hydrolysis}} \underset{\text{Phenol}}{ROH} + HCOONa$$

Thus,

$$\underset{\text{2,3,4-Trimethoxy benzaldehyde}}{\text{CHO; }OCH_3;\ OCH_3;\ OCH_3} \xrightarrow[\text{2) alc. KOH, }\Delta]{HCO_3H(0°)} \underset{\text{2,3,4-Trimethoxyphenol}}{\text{OH; }OCH_3;\ OCH_3;\ OCH_3}$$

Materials

Hydrogen peroxide (30%)	15 mL
Formic acid (98-100%)	75 mL
2,4,6-Trimethoxybenzaldehyde	5.0 g
Chloroform	50 mL
Sodium bisulphite	1.3 g

Procedure[1]

Step (i) Performic acid

Hydrogen peroxide (15 mL, 30%) is drpowise added to formic acid (75 mL, 98-100%). The mixture is warmed to 60° for 2 min. and then left at room temperature and stored at 0-5° in a refrigerator.

Step (ii) 2,3,4-Trimethoxyphenol

Performic solution (25 mL, 0.4 mole) is added dropwise to a stirred solution of 2,3, 4-trimethoxy-benzaldehyde (5 g, 0.025 mol) in dry chloroform (50 mL) kept at 0°. Addition of performic acid is stopped as soon as the absence of aldehyde is shown on TLC. The mixture is stirred continuously for 3 hr., treated with sodium bisulphite (1.3 g) and performic acid removed by distillation in vacuo. The residual solution is rendered alkaline by adding alcoholic potassium hydroxide solution (10%) and the mixture refluxed for 1 hr. with potassium hydroxide solution (10%, 50 mL). Alcohol is removed by distillation and the residue acidified with hydrochloric acid. It is extracted with ether (3 × 25 mL), ether extract dried (anhydrous sodium sulphate) and distilled to give 2,3,4-trimethoxyphenol as an oil yield 3.0 g (65%).

Note

1. V.K. Ahluwalli et al, Intermediates for organic synthesis, I.K. International, New Delhi, 2005, Page 27.
2. This is convenient method for the oxidation of methoxy aldehydes to phenols.
3. Using this procedure 3,4-dimethoxyphenol (from veratralde hyde) and 3,4-methylenedioxy phenol (from pipronal) were synthesised.
4. It is best to perform the reaction with polymer supported peracetic acid (See Section 11.7)

2.43 o-TOLUAMIDE

It is obtained by controlled hydrolysis of o-tolunitrile with alkaline hydrogen peroxide

$$2\ \text{o-Tolunitrile} + 2H_2O_2 \xrightarrow{^{\ominus}OH} 2\ \text{o-Toluamide} + O_2$$

o-Tolunitrile (structure: benzene ring with CH_3 and CN) → o-Toluamide (structure: benzene ring with CH_3 and $CONH_2$)

Materials

o-Tolunitrile	2.9 g
Hydrogen peroxide (30%)	10 mL
Sodium hydroxide solution	1 mL, 25%

Procedure

To a stirred solution of o-tolunitrile (2.9 g, 0.025 mol) in rectified spirit (12 mL) and sodium hydroxide solution (1 mL, 25%) is added slowly hydrogen peroxide (10 mL, 30%). The mixture becomes warm (about 40°) and kept at 40-50° (use ice-water cooling if necessary) The reaction takes about 1 hr. for completion. The temperature of the reaction mixture is kept at about 50° by external heating if necessary. The above solution (at 50°) is exactly neutralised (neutral to litmus) by addition of dilute H_2SO_4 (5%). The excess alcohol is removed at 20° under reduced pressure and the residual paste grinded with cold water (about 10 mL) in a mortar. The separated o-toluamide is fillered, washed with cold water and crystallised from hot water, m.p. 141°, yield 3 g (90%).

Notes

1. This method is useful for the controlled hydrolyses of both aromatic and aliphatic nitriles.
2. Though for o-tolunitrite, 30% H_2O_2 is used but for other nitriles 6-12% H_2O_2 gives satisfactory yields.
3. In order to minimise explosion danger, the reaction should be carried out in fume cupboard.
4. The acid amides can also be obtained form acid chlorides or esters by treatment with ammonia.
5. It can also be obtained by solid phase reaction of o-tolunitrile with urea-hydrogen peroxide (85°, 1.5 hr). For details see preparate of Benzamide (Section 3.4).

2.44 VANILLIDENEACETONE

It is obtained by the condensation of vanillin with acetone under basic conditions. The reaction is known as **Claisen-Schmidt reaction.**

O
‖
H — C(=O) + $H_2\bar{C}$ — C(=O) — CH_3 (ring: OCH_3, O^-) $\xrightarrow{\text{pH 14}}$ H — C($\ddot{O}^-$) — CH_2 — C(=O) — CH_3 (ring: OCH_3, O^-) (dianion) $\xrightarrow{H^+}$

H — C(OH) — C(H) — C(=O) — CH_3 (ring: OCH_3, O^-) H ← :$\bar{O}$H $\xrightarrow{H_2O}$ H — C = CH — C(=O) — CH_3 (ring: OCH_3, O)

$$\text{HC}{=}\text{CH}{-}\overset{\text{O}}{\overset{\|}{\text{C}}}{-}\text{CH}_3$$

(benzene ring bearing OCH_3 and OH substituents)

Vanilidene acetone

Materials

Vanillin	3.04 g
Acetone	12 mL
Sodium hydroxide solution (10%)	9 mL

Procedure

To a solution of vanillin (3.04 g) in acetone (12 mL) is added sodium hydroxide (10%, 9 mL). The reaction mixture is kept for 72 hrs. with occasional shaking. Water (5 mL) is added to the reaction mixture and then acidified with dilute hydrochloric acid (10%, 15 mL). The separated product is filtered, washed with water and crystallised form dilute alcohol. yield 3.3 g (85%). Its m.p. is recorded.

2.45 MISCELLANEOUS TRANSFORMATIONS IN WATER

Autocatalysis

The reactivity of some organic compounds in water in near critical region (250°C) or in high temperature (HTW) enhanced considerably (J.M. Kremsner and C.O. Kappe, Eur. J. Org. Chem., 2005, 3672). Under the above conditions, carboxylic acids produced by the hydrolysis of esters, aldehydes and amines and mineral acids (HX, HNO_3) produced by the hydrolysis of halogen and nitrogen compounds can act as acid catalyst. Similarly, ammonia produced by the hydrolysis of amines, amides and nitriles act as a basic catalyst. This is referred to as **autocatalysis**. Following are given the expected products obtained (N. Akiya and P.E. Savage, Chem. Rev. 2002, **107**, 2725 and the references cited there in)

$$\underset{\text{Ethers}}{\text{R}-\text{O}-\text{R}'} \xrightarrow{\text{HTW}} \text{ROH} + \text{O}-\text{R}'$$

$$\underset{\text{Esters}}{\text{R}-\text{COOR}'} \xrightarrow{\text{HTW}} \text{ROOH} + \text{R}'\text{OH}$$

$$\underset{\text{Amides}}{R—CONH_2} \xrightarrow{\text{HTW}} ROOH + NH_3$$

$$\underset{1^\circ\ \text{Amines}}{R—NH_2} \xrightarrow{\text{HTW}} ROH + NH_3$$

$$\underset{2^\circ\ \text{Amines}}{R—NHR'} \xrightarrow{\text{HTW}} ROH + R'NH_2$$

$$\underset{3^\circ\ \text{Amine}}{RR'—NR''} \xrightarrow{\text{HTW}} ROH + R'OH + R''NH_2$$

$$\underset{1^\circ\ \text{Nitroalkanes}}{RCHNO_2} \xrightarrow{\text{HTW}} RCHO + HNO_3$$

$$\underset{2^\circ\ \text{Nitroalkanes}}{RCNO_2R'} \xrightarrow{\text{HTW}} RCOR' + HNO_3$$

$$\underset{\text{Akyl halide}}{R\text{–}X} \xrightarrow{\text{HTW}} ROH + HX$$

$$\underset{1^\circ\ \text{gem-dihalides}}{RCX_2H} \xrightarrow{\text{HTW}} RCHO + 2HX$$

$$\underset{2^\circ\ \text{gem-dihalides}}{RCX_2R'} \xrightarrow{\text{HTW}} \underset{90\%}{RCOR' + 2HS}$$

$$\underset{R = \text{Alyl or aryl}}{RC{\equiv}N} \xrightarrow[H_2O]{280^\circ/6\ \text{min}} \overset{90\%}{\underset{90\%}{RCOOH}}$$

Some other interesting transformations with water at high temperature are given below (C.R. Strauss and K.W. Trainer, Aust. J. Chem., 1955, **48**, 1665 and the references cited therein).

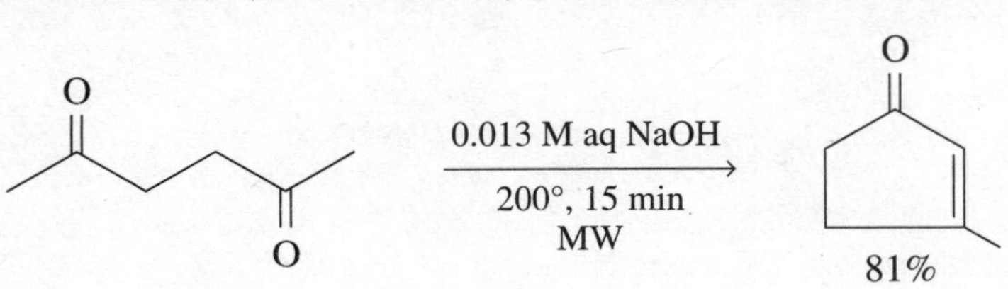

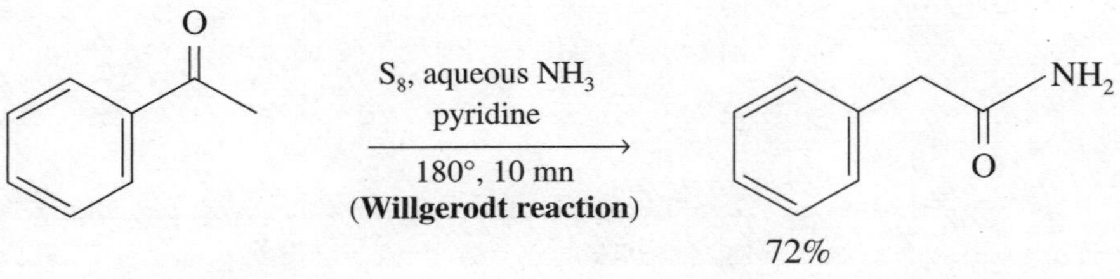

0.013 M aq NaOH
200°, 15 min
MW

81%

S_8, aqueous NH_3
pyridine
180°, 10 mn
(**Willgerodt reaction**)

NH_2

72%

2.46 CONCLUSION

Water is the best choice for organic transformations, since it is environmentally benign. The advantage of using water as a solvent is its cost, safety (since it is non-inflammable and is devoid of any carcinogenic effect) and simple operation. In fact water is used as a medium for promoting old and new reactions. A large number of reaction which were earlier performed using volatile organic solvents can be performed in water. According to CEN News (Sept 3, 2007), 'when organics fail, try water'. A number of organic reactions have been carried out in water. These includes Hofmann rearrangement, Benzil-Benzilic acid rearrangement, benzoin condensation, claisen-smith condensation, Dakins oxidation, Kolbe-Schmitt reaction, Baeyer-Villeger oxidation, pinacolic rearrangement, Diels-Alder reaction, Michael addition reaction, Aza. Diels-Alder reaction, Schotten-Baumann reaction, Baylis Hillman reaction, Haloform reaction, interamolecular aldol condensation, Sommelet Reaction, Pinacol-Pinacolone rearrangement, saponification, Wittig reaction etc.

Chapter 3

Transformations in Solid Phase

3.1 2-ALLYL PHENOL

It is obtained by **claisen rearrangement** of allyl phenyl ether. It is a [3,3]-sigmatropic rearrangement.

CH_2 O CH CH_2 $\xrightarrow{\Delta}$ O $\longrightarrow$ O H $CH_2 - CH = CH_2$

Allyl phenyl ether

Cyclic transition state

$\downarrow$

OH $CH_2 - CH = CH_2$

2-Allyl phenol

Materials

Allyl phenyl ether 2.1 g

Procedure

Allyl phenyl ether (2.1 g) is heated in a test tube using a sand bath and an air condenser. The liquid is gently refluxed for about 4 hr. The reaction mixture is cooled, sodium hydroxide solution added and extracted with ether. The clear alkaline solution is acidified (dil. HCl) and extracted with ether (2 × 15 mL). The ether extract is dried ($Na_2 SO_4$) and distilled to give 2-allyl phenol, b.p. 103-105°/19 mm. yield 1.5 g.

Notes

1. L. Claisen, Ber, 1912, **45**, 3157; L. Claisen and E. Tietze, Ber., 1925, **58**, 275; D.S. Tarbell, Org. React., 1944, **2**, 1.
2. Since it is a rearrangement reaction so there is 100% atom economy and so is a green reaction.
3. In claisen rearrangement, the allyl group migrates from oxygen to the ring preferably at ortho position. If both the o-positions are blocked, p-substituted phenol is obtained via two successive shifts of allyl group. An example is given below.

$O-CH_2CH=CH_2$ H_3C CH_3 $\xrightarrow{\Delta}$ O H_3C CH_3 CH_2 CH CH_2 * $\longrightarrow$ O H_3C CH_3 H $\overset{*}{C}H_2CH=CH_2$ $\longrightarrow$ $\longrightarrow$ OH H_3C $\underset{*}{C}H_2CH=CH_2$

4. The required allyl phenyl ether is obtained as follows: A mixture of phenol (1.9 g), allyl bromide (2.5 g), anhydrous potassium carbonate (3 g) and dry acetone (20 mL) is refluxed on a water bath using a round bottomed flask fitted with a reflux condenser and calcium chloride guard tube for 6 hr. The acetone is distilled (as much as possible) on a boiling water bath and water (15 mL) added to the residue. The mixture is extracted with ether (2 × 20 mL), extract washed with sodium hydroxide solution (10% 2 × 10 mL) and finally with water (2 × 10 mL). The ether solution is dried over anhydrous sodium sulphate and distilled. The allyl phenyl ether is obtained by distillation under reduced pressure b.p. 65°/19 mm, yield 2.1 g (78%)
5. Claisen rearrangement of allyl phenyl elter can also be conducted by heating in water at 240° (microwave oven) for 10 min. Alternatively, the reaction can be conducted in a scaled

tube at 240° (10 min heating). Yield 84% (K.D. Raner, C.R. Strauss and R.W. Trainor, J. Org. Chem., 1995, **60**, 2456; J. An, L Bagnell, T. Cablewski, C. R. Strauss and R.W. Trainor, J. Org. Chem., 1997, **62**, 2505.

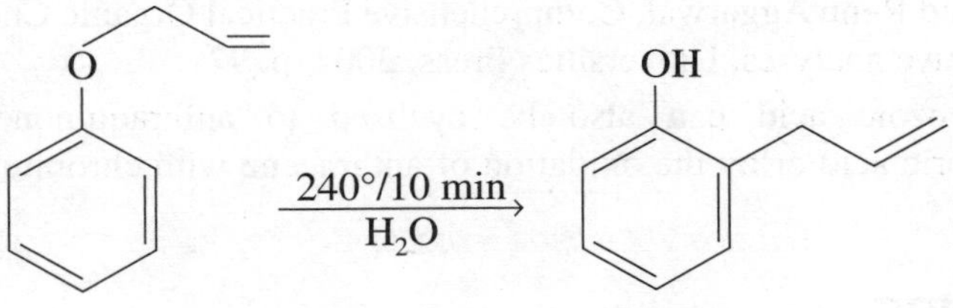

O OH

240°/10 min
H_2O

84%

However, if the above reaction is conducted at 258° for 60 min in water, 72% yield of 2- methyl-2,3-dihydrobenzofuran is obtained (C.R. Strauss, Aust. J. Chem., 1999, **52**, 86)

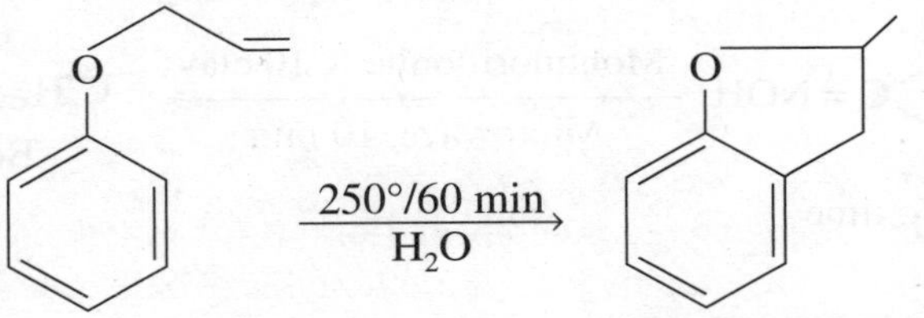

O O

250°/60 min
H_2O

72%

3.2 ANTHRAQUINONE

It is obtained by the cyclisation of o-benzoyl benzoicacid using microwaves in presence of acidic clay (G. Bram, A. Loupy, M. Majdoub and A. Petit, Chem., Ind., 1991, 396)

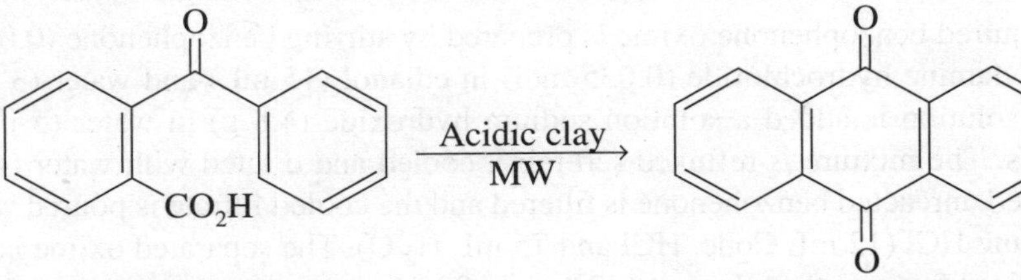

O O

Acidic clay
MW

CO_2H

O

o-Benzoyl benzoic acid Anthroquinine

Procedure

o-Benzoylbenzoic acid is mixed with acidic clay and the mixture irridated in presence of microwaves for 2 min. The formed anthraqumone is crystallised from glacial acetic acid, m.p. 285-286° yield 90%.

Notes

1. The required o-benzoylbenzoic acid is obtained by the Friedal-crafts reaction of benzoic acid with phthalic acid in presence of anhydrous aluminium chloride in benzene. (V.K. Ahluwalia and Renu Aggarwal, Comprehensive Practical Organic Chemistry: Preparation and quantitative analyses, Universities Press, 2004, p. 97)
2. o-Benzoylbenzoic acid can also be cyclised to anthraquinone by heating with polyphosphoric acid or by the oxidation of anthracene with chromium trioxide is glacial acetic acid.

3.3 BENZANILIDE

It is obtained by **Beckmann rearrangement** of benzophenone oxime with montmorillonite K 10 clay in dry media by irradiation with microwave.

$$(C_6H_5)_2C=NOH \xrightarrow[\text{Microwave, 10 min}]{\text{Montmorillonite K 10 clay}} C_6H_5-\overset{O}{\overset{\|}{C}}-NHC_6H_5$$

Benzophenoe oxime   Benzanilide

Procedure[1]

A solution of benzophenone oxime is added to montmorillonite K 10 clay, suspension stirred for 2-3 min and ether evaporated. The solid residue is heated in a microwave oven for 10 min. The formed benzanilide is crystallised from alcohol, m.p. 163-164°. yield 90%.

Notes

1. A. I. Bosch. P.de La Cruez, E. Diez-Barra, A. Loupy and F. Langa, Synlett., 1995, 1259.
2. The required benzophenone oxime is prepared by stirring benzophenone (0.022 mol) and hydroxylamine hydrochloride (0.035 mol) in ethanol (15 mL) and water (3 mL). To the stirred solution is added a solution sodium hydroxide (4.8 g) in water (5 mL) in small portions. The mixture is refluxed (20 min), cooled and diluted with water (40 mL). The separated unreacted benzohenone is filtered and the cooled filtrate is poured while stirring into dilute HCl (12 mL Conc. HCl and 75 mL H_2 O). The separated oxime is filtered and crystallised from methanol, m.p. 142°. yield 93%.
3. Beckmann rearrangement is usually performed by treatment of Ketoximes with acid catalyst (e.g., PCl_5, P_2O_5, $POCl_3$, H_2SO_4 etc), yield is approximately 80%.
4. Benzanilide can also be prepared by treating aniline with benzoyl chloride in presence of aqueous sodium hydroxide solution (**Schotten-Baumann Reaction**).
5. Use of acetophenone oxime in the above Beckmann rearrangement gives acetanilide, m.p. 114° in 90% yield.

3.4 BENZAMIDE

It is usually prepared from the corresponding carboxylic acid by treatment with thionyl chloride followed by treatment with ammonia. It is also obtained from the corresponding cyanide (C_6H_5CN) by hydrolysis under controlled conditions.

It is now possible to prepare benzamide in solid state by oxidation of cyanide with urea hydrogen peroxide complex (which is commercially available and can also be prepared)

$$\text{Benzonitrile } (C_6H_5CN) \xrightarrow[85°,\ 1.5\ \text{hr}]{\text{urea-hydrogen peroxide (UPH)}} \text{Benzamide } (C_6H_5CONH_2)\ (85\%)$$

Materials

Benzonitrile	2 g
Urea-hydrogen peroxide (complex) $[NH_2CONH_2 \cdot HOOH]$	3 g

Procedure[1]

Heat a mixture of benzonitrile (2 g) and urea-hydrogen peroxide complex (UPH) (3 g) in a glass tube at 85° (oil bath) for 1.5 hr. After completion of the reaction (as seen by TLC in the solvent system, hexane-ethylacetate, 8:2), extract the reaction mixture with ethyl acetate, dry the combined ethyl acetate extract over anhydrous sodium sulphate. Remove the solvent under reduced pressure and purify the product by chromatography. Benzamide, m.p. 130°, was obtained in 85% yield.

Notes

1. R. S. Varma and K.P. Naicker, Org. Letters., 1999, **1**, 189)
2. Using this procedures, phenylacetonitrite ($C_6H_5CH_2CN$) could be oxidised to phenyl acetamide ($C_6H_5CH_2CONH_2$) in 80% yield.

3.5 Benzil

Benzoin on oxidation with copper (II) sulfate-Al_2O_3[1] or oxone-wet alumina[2] under the influence of micro wave gives benzil

$$\underset{\text{Benzoin}}{C_6H_5CH(OH)COC_6H_5} \xrightarrow[\text{MW}]{CuSO_4 - Al_2O_3 \text{ or oxone - } AlO_3} \underset{\text{Benzil, } 88\%}{C_6H_5COCOC_6H_5}$$

The formed benzil in crystallised from alcohol, m.p. 95°, yield 88%

Notes

1. R.S. Varma, D.Kumar and R. Dahiya, J. Chem. Res(s) 1998, 324.
2. R.S. Varma, R. Dahiya and D. Kumar, Molecules on line, 1998, 2, 82
3. The starting benzoin is prepared by from benzaldehyde (see Section 2.4)
4. Oxone, a registered trade mark of E.I. du Pont de Nemours and company, is a mixture $2KHSO_5 . KHSO_4 . K_2SO_4$
5. Using the above procedure, a number of substitute benzoins could be oxidised into the corresponding benzils

$$R_1-CH(OH)-C(=O)-R_2 \xrightarrow[\text{MW}]{CuSO_4 - Al_2O_3 \text{ or oxone} - Al_2O_3} R_1-C(=O)-C(=O)-R_2$$

Benzoin — Benzil 88%

$R_1 = Me; R_2 = Ph$
$R_1 = R_2 = Ph = 4Me\ C_6H_4 = 4MeO\ C_6H_4$
$R_1 = Ph; R_2 = 4Me\ C_6H_4 = 4\text{-}MeO\ C_6H_4$

6. Benzil can also be obtained by the oxidation of benzoin (0.106 g) by heating a thoroughly mixed mixture of benzoin and clayfen (0.125 g) in a microwave oven for 60 see. The yield in 94%. (R.S. Varma, Clean Products and Processes, 1999, P.139)

$$C_6H_5-CH(OH)-C(=O)-C_6H_5 \xrightarrow[\text{MW, 60 sec}]{\text{Clayfen}} C_6H_5CO-CO-C_6H_5 \quad 94\ \%$$

Benzoin

Oxidation of benzil to benzoin can also be effected by heating with 35% MnO_2 'doped' silica and MW irradiation conditions.

$$C_6H_5-CH(OH)-C(=O)-C_6H_5 \xrightarrow[\text{M.W, 45 sec}]{MnO_2 - \text{silica}} C_6H_5CO-CO-C_6H_5 \quad 85\%$$

Another procedure of oxidation of benzil to benzoin is by heating a mixture of benzoin and CrO_3 imprignated on pre-moistened alumina (in the ratio 1:3) in a MW oven for 30 sec. yield 70% (R.S. Verma and R.K. Saini, Tetrahedron Lett., 1948, **39**, 1481)

$$C_6H_5-CH(OH)-C(=O)-C_6H_5 \xrightarrow[\text{M.W, 30 sec}]{\text{Wet } CrO_3 - Al_2O_3} C_6H_5COCOC_6H_5 \quad 70\%$$

Benzoin can also be oxidised by using iodobenzene diacetate (IBD) as an oxidant under solvent free condition using MW irradiation (R.S. Verma, R. Dahiya and R.K. Saini, Tetrahedron Lett., 1998, **38**, 7029)

$$C_6H_5\underset{\substack{|\\OH}}{CH}-\underset{\substack{\|\\O}}{C}-C_6H_5 \xrightarrow[\text{M.W, 2 min}]{\text{IBD/Alumina}} \underset{90\ \%}{C_6H_5CO\ CO\ C_6H_5}$$

3.5a BENZONITRILE

A single step conversion of benzaldehyde to benzonitrile has been developed. The procedure[1] involves in heating benzaldehyde with clay supported hydroxylamine hydrochloride and coupled with MW irradiation in absence of solvent.

$$C_6H_5CHO \xrightarrow[\text{MW, 1.5 min}]{\text{K10 clay-}NH_2OH\cdot HCl} \underset{92\%}{C_6H_5CN}$$

Benzonitrile was isolated by extraction with methylene chloride and final distillation, b.p. 197° [b.p. 123.5°/100 mm pr] in 92% yield

Notes

1. R.S. Varma, K.P. Naicker, D. Kumar and R. Dahiya Molecules, J. Microwave Power Electromagn. Energy, 1999, **34**, 113.
2. Using the above procedure a number of substituted benzonitriles could be prepared from appropriately substituted benzaldehydes

$$R_1,R_2\text{-}C_6H_3\text{-}CHO \xrightarrow[\text{MW, 1-2 min}]{\text{K10 clay-}NH_2OH\cdot HCl} \underset{90\text{-}95\%}{R_1,R_2\text{-}C_6H_3\text{-}C\equiv N}$$

$R_1 = R_2 = H = OMe$

$R_1 = H;\ R_2 = OH, Br, Me, OMe, NO_2$

3. Benzonitrile can also be prepared form benzamide by treatment with chloroform and aqueous sodium hydroxide (stirring 2 hr. at room temp.) in presence of benzyltriethyl ammonium chloride (T. Saraie, K. Ishiguro, K. Kawashina and K. Morila, Tetrahedron Lett., 1973, 2121). The reaction involved is

$$C_6H_5-\overset{O}{\overset{||}{C}}-N\begin{matrix}H\\H\end{matrix} + :C\begin{matrix}Cl\\Cl\end{matrix} \longrightarrow C_6H_5-\overset{O-\bar{C}(Cl)_2}{\overset{|}{C}}-\overset{+}{N}H_2 \longrightarrow C_6H_5C\begin{matrix}O-CH(Cl)_2\\=N-H\end{matrix} \xrightarrow{^-OH} C_6H_5C\equiv N$$

Benzamide dichlorocarbene (generaled in situ)

5. The usual method of converting an aldooxime into the corresponding nitrite is by dehydration on heating with P_2O_5

3.6 BENZYL ALCOHOL

Aldehydes can be reduced to alcohols by using barium hydroxide $[Ba(OH)_2.8H_2O]$ and Paraformaldehyde. Thus the reaction of benzaldehyde and paraformaldehyde in presence of barium hydroxide (MW heating) gives benzyl alcohol. Thus is an illustration of **solid state cannizzaro reaction**.

$$C_6H_5CHO + (CH_2O)_2 \xrightarrow[MW]{Ba(OH)_2\cdot 8H_2O} C_6H_5\ C_6H_5OH + C_6H_5COOH$$

Benzaldehyde Para formaldehyde — 91% (1-20%) Benzyl alcohol

Materials

Benzaldehyde	1.0 g
Paraformaldehyde	6.0 g
Barium hydroxides octahydrate	6.0 g

Procedure[1]

Add benzaldehyde (1 g) to finely powdered parformaldehyde (6 g). To this mixture add powdered barium hydroxide octahydrate (6 g). The above mixture contained in a glass tube was placed in alumina bath (natural alumina, 200 g, mesh-150, Aldrich, bath about 10 cm diameter) inside a household microwave oven and irradiated for about 10 min. at its full Power of 900 watts. After completion of the reaction [as indicated by TLC (hexane-EtOAc, 4:1, v/v)] extract with ethyl acetate. Dry the combined ethyl acetate extract over anhydrous sodium sulphate and distil. Benzyl alcohol (b.p. 204°) was obtained in 91% yield. It can be further purified by distillation in vacuum. b.p. 185°/400 at. pr.

Notes

1. R.S. Varma, K.P. Naicker and P.J. Liesen, Tetrahedron Lett., 1998, 39, 8437
2. This reaction is known as **Crossed Cannizzaro Reaction**.

3. The above procedure can be used for the preparation of primary alcohols from aldehydes.
4. The above reaction can also be performed by heating in an oil bath (110-110°)
5. Benzyl alcohol can also be prepared by **crossed cannizzaro reaction** between benzaldehyde and formaldehyde in presence of methanolic potassium hydroxide. However, the above procedure is very much convenient.
6. Benzyl alcohol can be prepared by the reduction of benzaldehyde using Ru HCl (CO) (PPh_3) in presence of formic acid under microwave irradiation (L. Loupsy, P. Pigeon, M. Ramdani and P. Jalquault, Synthetic Commun., 1994, **24**, 159)

$$C_6H_5CHO + HCO_2H \xrightarrow[\text{RuHCl (CO) (PPh}_3)_3]{\text{MW, 7 min}} C_6H_5CH_2OH$$

3.7 BENZYLIDENE ANILINE

Benzylidene aniline, a Schiff's base or an imine is generally prepared from aromatic amines and aldehydes using several reagents like zinc chloride, titanium (IV) chloride, molecular sieves and alumina. However, the best procedure is to heat the aldehyde and amine in microwave oven under solvent free conditions using montmorillonite K Clay

$$\underset{\text{Aniline}}{C_6H_5NH_2} + \underset{\text{Benzaldehyde}}{OH\,C\,C_6H_5} \xrightarrow[\text{K-10, Clay, 60 sec.}]{\text{MW}} \left[C_6H_5\overset{\overset{\displaystyle H}{|}}{\underset{\underset{\displaystyle OH}{|}}{C}} - NH - C_6H_5 \right]$$

$$\xrightarrow{-H_2O} \underset{\text{Benzylidene aniline}}{C_6H_5CH{=}N - C_6H_5}$$

Both aniline and benzaldehyde are taken in 1:1 ratio and heated in MW for 60 sec in presence of montmorillonite K-Clay. The yield is about 98%.

Notes

1. S.K. Dewan, U. Varma and S.D. Malik, J. Chem, Res.(S), 1995, 21; R.S. Varma, Tetrahedron, 2002, **58**, 1243 and references cited there in
2. Using this procedure a number of imines or Schiff's bases have been prepared

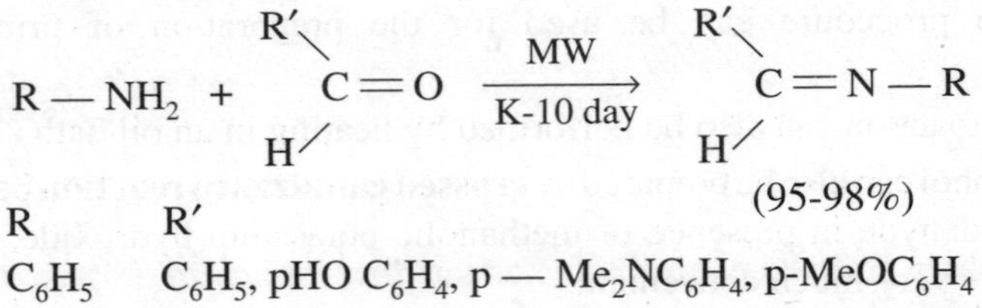

$$R-NH_2 + \underset{H}{\overset{R'}{>}}C=O \xrightarrow[\text{K-10 day}]{\text{MW}} \underset{H}{\overset{R'}{>}}C=N-R$$

(95-98%)

R	R′
C_6H_5	C_6H_5, pHO C_6H_4, p — $Me_2NC_6H_4$, p-MeOC_6H_4

3.8 2-CARBETHOXYCYCLOPENTANONE

It is obtained[1] by heating a mixture of diethyl adipate with sodium ethoxide. The reaction products are obtained by neutralising the cooled reaction with calculated amount of p-TSOH. H_2O.

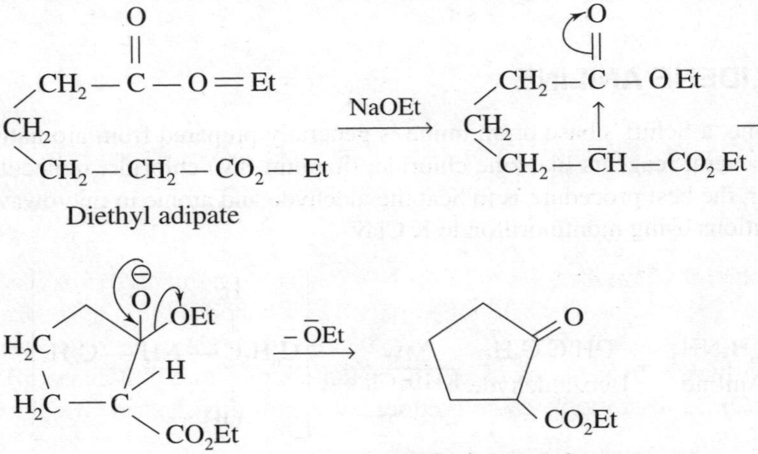

$$CH_2(CH_2-CO-OEt)(CH_2-CH_2-CO_2-Et) \xrightarrow{\text{NaOEt}} CH_2(CH_2-C(=O)-OEt)(CH_2-\overline{C}H-CO_2Et) \longrightarrow$$

Diethyl adipate

$$\xrightarrow{-\overline{O}Et}$$

2 - Carbethoxycyclopentanane
90%

The product, 2-carbethoxycyclopentanone was obtained, b.p. 108-111°(15 mm) in 90% yield. This is a case of **intramolecular claisen condensation**, known as **Dieckmann condensation**[2]

Notes

1. V.K. Ahluwalia. Green Chemistry, Environmentally Benign Reaction, Ane Books, India, 2006, P.83.
2. W. Dieckmann, Ber., 1894, **27**, 102, 965; 1900, 33, 2670
3. Using diethyl pimelate, 2-carbethoxycyclohexanone is obtained
4. Dieckmann condensation also proceeds very well on sonication (J.L. Luche, C. Petrier and C. Duputy, Tetrahedron Lett., 1985, **26**, 753.

$$EtO_2C - (CH_2)_4\, CO_2Et \xrightarrow[\text{toluene, 5 min}]{\text{K,))))}} \text{2-Carbethoxycyclopentanone (90\%)}$$

Diethyl adipate

2-Carbethoxycyclopentanone

90%

3.9 CATECHOL

It is best prepared by the oxidation of o-hydroxybenzaldehyde using urea-hydrogen peroxide complex usually called urea-hydrogen peroxide (UPH), which is commercially available and can also be easily prepared. The oxidation is conducted in solid state by heating the mixture of the aldehyde and UPH for 20 min. (at 85°).

urea-hydrogen peroxide (UPH)

$H_2N\,CONH_2$

|

HOOH

85°, 20 min

o-Hydroxy benzaldehyde → Catechol

80%

Materials

o-Hydroxybenzaldehyde	6.1 g
Urea-Hydrogenperoxide (UPH)	10.6 g

Procedure[1]

A mixture of o-hydroxybenzaldehyde (6.1 g) and urea-hydrogen peroxide (UPH) (10.6 g, 0.05 mol) is heated in an oil bath for 20 min at 85°. After the completion of the reaction (as seen by TLC in the solvent system hexane-ethylacetate, 8:2)], the reaction mixture is extracted with ethyl acetate. The extract is dried (anhydrous sodium sulphate and solvent removed under reduced pressure and residual product purified by column chromatography. Catechol, m.p. 136-138, obtained in 80% yield (4.2g).

Notes

1. R.S. Varma and K.P. Naicker, Organic Letters. 1999, **1**, 189
2. The oxidation of aldehydes to phenols by UPH in the solid state is a very convenient procedure compared to oxidation with alkaline hydrogen peroxide (Dakins oxidation) (see Section 2.11 or peracetic acid (Baeyer-Villiger oxidation) (see 2.42).

3. Using the above procedure a number of aldehydes like p-hydroxybenzaldehyde, 2-hydroxy-4-nitrobenzaldehyde and p-methoxybenzaldehyde and ketones like o-and p-hydroxyacetophenones have been oxidized[1] to the corresponding hydroxycompounds in good yields.

3.10 1,4-DIHYDRO-QUINOXALINE-2,3-DIONE

It is prepared by grinding together o-phenylene diamine and oxalic acid dihydrate.

o-Phenylene diamine + Oxalic acid $\xrightarrow[\text{RT}]{\text{Grinding}}$ 1, 4-Dihydroquinoxaline-2, 3-dione

Materials

o-Phenylene diamine	0.25 g
Oxalic acid dihydrate	0.21 g

Procedure

Grind a mixture of o-phenylene diamine (0.25 g) and oxalic acid dihydrate (0.21 g) in a pestle-mortar at room temperature until the mixture turned into a melt. Continue grinding intermittantly for 30 min. Crystallise the product from water or dilute alcohol yield 98%

Notes

1. H. Thakuria and G. Das, J.Chem. Sci, 2006, **118**, 425.
2. Different o-phenylenediamines could be used in this preparation. The subsitutents (NO_2, Cl, Me, n-Pr, Ph) can be present in position 4.
3. The 1,4-dihydroquinoxaline 2,3-diones are widely used in many fields, as curative intermediate, bacteriocides and insecticides.
4. In ths reaction there is 100% atom economy.

3.11 5,5-DIMETHYL HYDANTOIN

It is obtained by heating cyanohydrin of acetone with ammonium carbonate. The reaction is know as **Bucherer hydantoin synthesis**[1].

$$\underset{\text{Acetone cyanohydrin}}{(H_3C)_2C(CN)(OH)} \xrightarrow[\Delta]{(NH_4)_2CO_3} \underset{\text{5, 5-Dimethylhydantoin}}{\text{5,5-dimethylhydantoin}}$$

Procedure

A mixture of acetone cyanohydrin (7.01 g, 0.88 mol) and powdered ammonium carbonate (12 g) is warmed on a water bath (fume cupboard). The mixture is stirred and heated for 2 hr at 70-80° and finally for 30 min at 90° and cooled. The separated 5, 5-dimethylhydantoin is crystallised from hot water, m.p. 178°. Yield 4.25 g (44.6%)

Notes

1. H.T. Bucherer and H.T. Fischbeck, J. Prakt. Chem., 1934, **140**, 19; H.T. Bucherer and W. Sleiner, J. Prakt. Chem., 1934, **140**, 291.
2. Acetone cyanohydoin is obtained as follows. Acetone (14.25 mL, 0.2 mol) is added to a solution of sodium metabisulphite (11 g) in water (20 mL) in a R.B. flask. A solution of KCN (6 g) in cold water (20 mL) is added slowly with stirring. The upper layer of acetone cyanohydrin is separated (separatory funnel), dried (Na_2SO_4) and solvent removed. Yield 7.5 g (89%), b.p. 80-82°/15 mm.
3. The Bucherer hydantoin synthesis can also be conducted by heating in a domestic microwave oven for 30 sec.

3.12 2,5-DIMETHYLPYRROLE

It is obtained by the condensation of hexane-2,5-dione with ammonium carbonate. The reaction is known as **Paal-knorr synthesis**.

$$\underset{\text{Hexane-2,5-dione}}{CH_3\cdot CO{-}CH_2{-}CH_2{-}CO\cdot CH_3} + (NH_4)_2CO_3 \xrightarrow{\Delta,\ 100^\circ} \underset{\text{2,5-Dimethylpyrrole}}{\text{2,5-dimethylpyrrole}}$$

Materials

Hexane-2,5-dione	5 g
Ammonium carbonate	10 g

Procedure

A mixture of hexane-2,5-dione (5 g, 5.15 mL, 0.044 mol) and ammonium carbonate (10 g, excess) is heated at 100° (oil-bath) until etterenscene ceases (use an small air-condenses) (60-70 min). To the cooled mixture, is added water (2 mL), extract with chloroform (2 × 10 mL), dry the extract (Na_2SO_4) and distil. 2,5-Dimethyl pyrrole, bp. 78-80°/20 mm is obtained in 86% yield (3.6 g).

Notes

1. The reaction can also be conducted in a micro wave over (5 min heating) is a beaker covered with a watch glass.
2. The required hexane-2,5-dione is obtained from ethyl acetoacetate as give below.

$$2CH_3COCH_2CO_2Et \xrightarrow{2Na} 2\left[CH_3COCHCO_2Et\right]^{-} \overset{+}{Na}$$

$$\xrightarrow{I_2} \begin{array}{c} CH_3COCHCO_2Et \\ | \\ CH_3COCHCO_2Et \end{array} \xrightarrow[\text{decarboxylation}]{\text{Hydrolysis}} CO_3COCH_2COCOCH_3$$

3.13 DIPHENYLCARBINOL

It is obtained by the reduction of benzophenone with sodium borohydride in solid state.

$$\underset{\text{Benzophenone}}{Ph_2CO} + \underset{\text{Sodium borohydride}}{NaBH_4} \longrightarrow \underset{\text{Diphenylcarbinol}}{Ph_2CHOH}$$

Materials

Benzophenone	1.8 g
Sodium borohydride	4.8 g

Procedure[1]

An intimate mixture of benzophenone (1.8 g) and sodium borohydride (4.8 g) is kept is a dark box at room temperature with occasional mixing and grinding for 5 days. The product obtained is diphenyl carbinol, yield 100%.

Notes

1. K. Tanka and F. Toda, Chem. Rev., 2000, **100**, 1028.

3.14 Flavone

Flavones have been synthesised by various methods such as Allan-Robinson Synthesis. The most popular route involves the Baker-Venkataraman rearrangement where in o-hydroxyacetophenone is benzoylated followed by the base (pyridine/KOH) treatment of the benzoyl ester to effect an acyl group migration to produce 1,3-diketone. The cyclisation of the diketone is usually accomplished under strongly acidic conditions using sulfuric acid and acetic acid. A facile and clean approach for the cyclization step utilizes benign and readily available starting materials. Thus, o-hydroydibenzoyl methane intermediate was adsorbed on montmorillonite K 10 clay and irradiated with micro wave for 1-1.5 min to give the flavone.

OH O O → K10 clay, MW, 1-1.6 min → O O

o-Hydroxy dibenzoyl methane

Flavone m.p 99-100°

Procedure

Take o-Hydrxydibenzoyl methane (0.2 g) in a glass tube, dissolved in small amount of dichloromethane (1 mL) and adsorbed on montm orillonite K-10 clay (1 g). Place the test tube in an alumina bath inside the microwave oven and irradiate for 1.5 min. Extract the crude product into dichloromethane (2 × 15 mL). Evaporate the solvent and crystallise the product from methanol. M.P 99-100° yield 75%.

Notes

1. R.S. Varma, R.K. Saini and D. Kumar, J. Chem. Res (S), 1998, 3400
2. The starting materials. o-hydroxydibenzoyl methane is prepared as given in Section 5.9
3. Using the above procedure a number of flavones were prepared.

OH R_1 R O O → K10 Clay, MW, 1-1.5min → O R_1 R O

75-80%

$R = H; R_1 = H, p\text{-}CH_3, p — OCH_3, p — NO_2$
$R = OCH_3; R_1 = H, p\text{-}CH_3, p — CH_3$

3.15 IMIDAZO [1,2-a] PYRIDINES

These are obtained from aldehydes, 2-aminopyridine anal isonitriles in presence of catalytic amount of clay by heating in a microwave oven.

$C_6H_5CHOC_6H_5$ + 2-Amino-pyridine (NH_2, N) + $C_6H_5CH_2NC$ $\xrightarrow[\text{clay, 1 min}]{\text{MW}}$ Imidazolo [1, 2-a] pyridines (C_6H_5, N, $C_6H_5CH_2N$ H, N)

Benzaldehyde

Benzyl Isonitrile

Procedure[1]

A mixture of benzaldehyde (106 mg, 1 m mole) and 2-aminopyridine (94 mg, 1 m mole) was irradiated in a microwave oven for 1 min (at full power of 900 W) in the presence of montmorillonite K-10 clay (5 mg). After the addition of benzyl isocyanide (117 mg, 1 m mol), the reaction mixture was further irradiated successively (2 min) at 50% power level for a duration of 1 minute followed by cooling (1 minute). The formed product was dissolved in dichloromethane (2 × 5 mL), fillered and solvent distilled. The crude product was purified by crystallisation or by passing through a small bed of silica gel using ethyl acetate: hexane (4: 1, v/v) as eiuant to obtain the substituted imidazo [1, 2-a] pyridine, m.p 112-113°.

Notes

1. R.S. Varma and D. Kumar, Tetrahedron Lett., 1999, **40**, 7665.
2. The condensation of aldehydes with isocyanides and 2-amino pyridine is known as **ugi reaction.**
3. The above procedure (ugi reaction) can also be used for the synthesis of imidazolo [1,2-a] pyrimidines and imidazoto [1,2-a] pyrimidines by using 2-aminopyrazine and 2-aminopyridimes respectively in place of 2-aminopyridines.
4. The above process is general for all the three components, e.g., aldehydes (aliphatic, aromatic and vinylic), isocyanides (aliphatic aromatic and cyclic) and amines (2-amino pyridine, 2-aminopyrazine and 2-aminopyrimidine).

3.16 3-NITRO-2-PHENYL-2H-CHROMENES

It is obtained by the condensation of o-hydroxyl benzaldehyde with ω-nitrostyrene in presence of basic alumina using sonication.

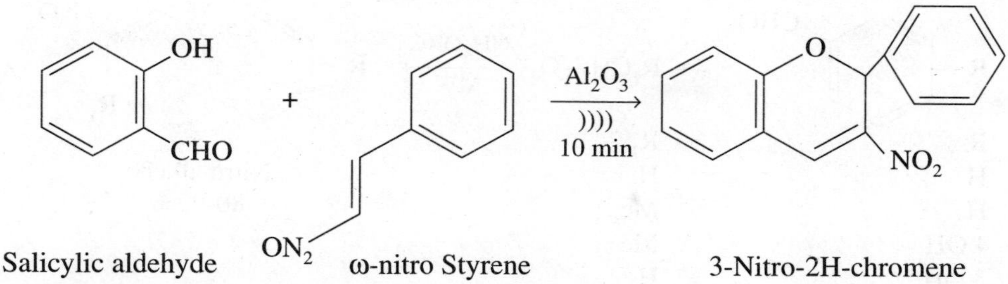

OH CHO + ON$_2$ $\xrightarrow[\text{10 min}]{Al_2O_3 \text{))))}}$ O NO$_2$

Salicylic aldehyde ω-nitro Styrene 3-Nitro-2H-chromene

Procedure[1]

A mixture of salicylic aldehyde (0.02 mole) and ω-nitrostyrene (0.02 mol) was absorbed on basic alumina (30 g) using methylone chloride (15 mL). The solid material on sonication using low intensity untrasonic bath for 10 min gave 3-Nitro-2-phenyl 2H-chromene in 80% yield.

Notes

1. R.S. Varma and G.W. Kabalka, Heterocycles, 1985, **23**, 189.

3.17 ω-NITROSTYRENE

It is prepared by the condensation of benzaldehyde with nitromethane (in presence of catalytic amount of ammonium acetate) in microwave oven.

$$\underset{\text{Benzaldehyde}}{C_6H_5CHO} + \underset{\text{Nitro methane}}{CH_3NO_2} \xrightarrow[\text{MW, 8 min}]{NH_4OAc} \underset{\text{ω-Nitrostyrene 80\%}}{C_6H_5CH{=}CHNO_2}$$

Procedure

Heating a mixture of benzalddehyde and nitro methane (1: 1) in presence of catalytic amount of ammonium acetate in a microwave oven for 8 min gives 80% yield of ω-nitrostyrene, m.p. 58-59°.

Notes

1. R.S. Varma and R. Dahiya, Tetrahedron lett., 1997, **38**, 2043
2. The above reaction involving condensation of carbonyl compounds with nitralkanes to afford nitro alkenes is known as **Henry reaction**
3. Using the above procedure a number of nitro alkenes can be obtained

R–C6H4–CHO + $R_1CH_2NO_2$ $\xrightarrow[MW]{NH_4OAC}$ R–C6H4–CH=C(R_1)NO_2

R	R_1
H	H
H	Me
4 OH	Me
4 OH	H
3,4-$(MeO)_2$	H or Me

Nitro alkene
80-92%

4. ω-Nitrostyrene can also be obtained by nitration of styrene with clay 'doped' with nitrate salts by heating in a bath (100-110°, 15 min) or irradiated in miwave oven (100-110°, 3 min) (R.S. Varma, K.P. Naricker and P.J. Liesen, Tetrahedron Left, 1998, **39**, 3977)

Styrene $\xrightarrow[\text{oil bath or MV, 110-110°}]{\text{Clayfen (Clayan)}}$ X–C6H4–CH=CH–NO_2 (14-56%) + X–C6H4–CHO (9-21%)

Styrene
p-chlorostyrene, X = Cl
p-methylstyrene, X = Me
p-methoxy styrene, X = OMe

The ω-nitrostyrene is obtained in 14-56% yield A, byproduct (corresponding benzaldehyde) is also obtained in minor amounts (9-21%)

5. ω-Nitrostyrene can also be obtained by the condensation of benzaldehyde with nitromethane in presence of alkali. Thus, benzaldehyde (5.85 g, 5.6 ml, 0.055 mol) and nitromethane (3.75 g, 3.32 mL, 0.061 mol) are dissoled in methanol (20 mL). The solution is cooled (freezing mixture) and to the stirried solution is added a solution of NaOH (2.4 g) in water (6 ml); the temperature is kept between 10-15°. The mixture is stirred for 15 min, cold water (35 mL) added followed by addition of dilute HCl (12.5 mL conc. HCl in 19 mL H_2O). The separated nitromethare is filtered and crystallsid from alcohol. M.P 57-58°, yield 6.25 g (76.2%).

3.18 SEC.PHENETHYL ALCOHOL

It is prepared by the reduction of acetophenone with alumina-supported $NaBH_4$ under microwave irradiation

$$C_6H_5-\overset{\overset{\displaystyle O}{\|}}{C}-CH_3 \xrightarrow[\text{MW, 30 Sec.}]{NaBH_4-Al_2O_3} C_6H_5-\overset{\overset{\displaystyle OH}{|}}{CH}-CH_3$$

Acetophenone → Sec. Phenethylalcohol 87%

Procedure[1]

Place a freshly prepared mixture of $NaBH_4$-alumina (1.13 g, 3.0 m mol of $NaBH_4$), pure acetophenone (0.36 g, 3.0 m mol) is a test tube and place it is an alumina bath inside a household oven (operating at 2450 MHz) and irradiated for 30 sec. After completion of the reaction (as monitored by TLC (hexane – EtOAc, 8:2, v\v), extract the product into methylene chloride. Remove the solvent. Pure sec phenethyl alcohol is obtained in 87% yield.

Notes

1. R.S. Varma, R. Dahiya and R.K. Saini, Tetrahedron Left, 1997, 38, 8819; R.S. Varma, Clean Products and processes 1999, **1**, 132-147 and reference cited there in.
2. 10% $NaBH_4$-Alumina was prepared by thoroughly mixing $NaBH_4$ (5 g) with neutral alumina (45 g) is solid state using pestle mortar; admixing the three components, carbonyl substrate, $NaBH_4$ and alumina together was equally efficient. The use of premoistened alumina accelerated further the reaction.
3. Using the above procedure a number of carbonyl compounds could be reduced.

$$R_1-C_6H_4-\overset{\overset{\displaystyle O}{\|}}{C}-R_2 \xrightarrow[\text{Micro Wave, 30-120 Sec.}]{NaBH_4-Al_2O_3} R_1-C_6H_4-\overset{\overset{\displaystyle OH}{|}}{C}-R_2$$

62-92%

$R_1 = H$; $R_2 = Me, Ph, -CH(OH)Ph$
$R_1 = Cl$; $R_2 = H$
$R_1 = Me$, $R_2 = H, Me$
$R_1 = NO_2$; $R_2 = H$

4. Acetophenone could also be reduced enzymatically (using Deucus carota roots) to 1-Phenyl- (1 S) ethan-1-ol in 73% yield (with ee 92). For details see Section. 8.4.
5. In PEG-400, carbonyl compounds can be reduced by $NaBH_4$ (E. Santaniello, A. Marzocchi and P. Sozzani, Tetrahedron Lett, 1999, **20**, 4581).

3.19 PHENYL BENZOATE

It is obtained by the **Baeyer-Villiger Oxidation** of bezophenone with in-chloro perbenzoic acid in solid state at room temperature.

$$\underset{\text{Benzophenone}}{C_6H_5COC_6H_5} \xrightarrow{\text{m-Chloroperbenzoic acid}} \underset{\text{Phenyl benzoate}}{C_6H_5COOC_6H_5}$$

Materials

Benzophenone	1.8 g
m-Chloroperbenzoic acid	3.5 g

Procedure[1]

An intimate mixture of benzophenone (1.8 g) and m-chloroperbenzoic and (3.5 g) is kept at room temperature for 24 hr. The formed produced is macerated with sodium bicarbonate solution and extracted with ether. The ether extract is dried (Na_2SO_4) and distilled to give phenyl benzoate in 85% yield B.P. 314°.

Notes

1. K. Tank and F. Toda, Chem. Rev., 2000, **100**, 1028
2. See also preparation of phenyl acetate, Section 2.34.
3. The conventional Baeyer-villiger oxidation in chloroform gives low yields (40-50%) of the product.

3.20 2-PHENYL-1,2,3,4-TETRAHYDRO-4-QUINOLONE

It is obtained by irradiation of o-aminochalcone.

$$\text{o-Amino chalcone} \xrightarrow[\text{MW, 1-2 min}]{\text{K-10, Clay}} \underset{80\%}{\text{2-Phenyl-1,2,3,4-tetrahydro-4-quinolone}}$$

Procedure

Mix o-aminochalcone (0.1 g) with montmorillonite K-10 clay (1 g) in solid state using a pestle and mortar. Transfer the mixture into a glass tube and place in an alumina bath (alumina, 100 g. mess

65-325, Fisher scientific, bath 6-8 cm diameter) inside the microwave oven. Irridate the mixture for 1.5 min (The temperature of the bath reached 100°). Completion of the reaction was inferred by TLC examination. Extract the product with dichloromethane (2 × 15 mL), filter and evaporate. 2-Phenyl-1, 2, 3, 4-tetrahydro-4-quinolone. mp 148-150° was obtained in 85% yield.

Notes

1. R.S. Varma and R.K. Saini, Synlet, 1997, 857
2. The starting o-amino chalcones were obtained from o-amino acetophenone by condensation with benzaldehyde in presence the sodium hydroxide.
3. Using the above procedure a number of 2-Phenyl-1, 2, 3, 4-tetrahydro-4-quinolones were obtained.

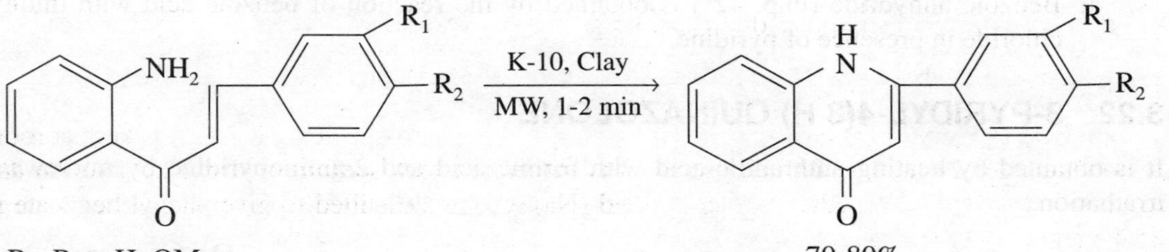

NH_2 ... R_1, R_2 $\xrightarrow[\text{MW, 1-2 min}]{\text{K-10, Clay}}$ H N ... R_1, R_2 (O) 70-80%

R_1, R_2 = H, OMe
R_1 = Me; R_2 = H
R_1 = OMe; R_2 = H
R_1 = Cl, Br, NO_2; R_2 = H

3.21 PHTHALIC ANHYDRIDE

It is obtained by sublimation of phthalic acid.

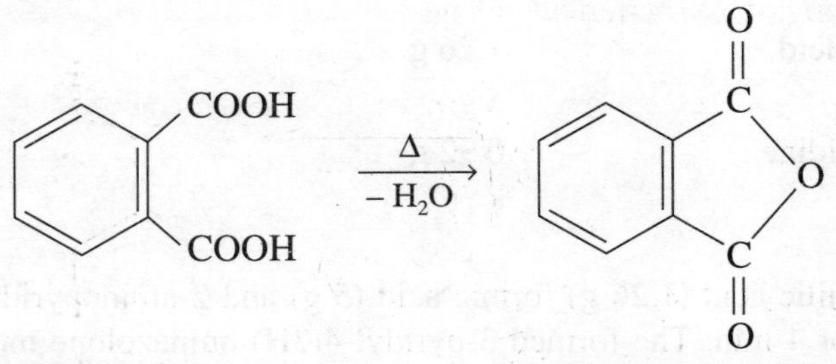

COOH, COOH $\xrightarrow[-H_2O]{\Delta}$ O=C–O–C=O

Phthalic acid — Phthalic anhydride

Procedure

Phthalic acid (5 g, 0.03 mol) is placed in a porcelin dish (2½ cm diameter), and a circular filter paper having circular holes is placed on the dish. The filter paper is covered by an inverted water-

Jacked glass funnel whose stem is plugged loosely with cotton or filter paper An ordinary inverted funnel can also be used. The dish is heated on a sand bath with the help of a flame. Phthalic anhydride formed rises through the holes in the filter paper and condenses in the hollow of the funnel on the filter paper, yield 4 g (89.9%) M.P. 132°.

Notes

1. Phthalic anhydrides can also be obtained by heating phthalic and with acetic anhydride.
2. Succinic anhydride (m.p. 119-120°) is obtained by heating succinic and with acetic anhydride
3. Maleic anhydride (m.p 54°) is obtained by distilling maleic acid in an inert high boiling solvent.
4. Benzoic anhydride (m.p. 42°) is obtained by the reaction of benzoic acid with thionyl chloride in presence of pyridine.

3.22 3-PYRIDYL-4(3 H) QUINAZOLONE

It is obtained by heating anthranilic-acid with formic acid and 2-aminopyridine by microwave irradiation.

COOH, NH_2 + HCOOH + NH_2, N —Micro wave→ O, N, N, N

Anthranilic acid Formic acid 2-Amino pyridine 3-Pyridyl-4(3H) quinazolone

Materials

Anthranilic acid	1.26 g
Formic acid	5 g
2-Aminopyridine	0.92 g

Procedure[1]

A mixture of anthranilic acid (1.26 g) formic acid (5 g) and 2-aminopyridine (0.92) is heated in a microwave oven for 4 min. The formed 3-pyridyl-4(3H) quinazolone melted at 156-57°, yield 92%.

References

1. M. Kidwai, S. Rastogi, R. Mohan and R. Ruby, Chemica Acta, 2003, **76(4)** 365

3.23 CIS-1, 2, 3, 6-TETRAHYDRO-4, 5-DIMETHYL PHTHALIC ANHYDRIDE

It is obtained by **Diels-Alder reaction** of 2, 3-dimethylbuta-1, 3-diene with maleic anhydride.

2,3-Dimethyl buta-1,3-diene + Maleic anhydride ⟶ cis-1,2,3,6-Tetrahydrio-4,5, -dimethylphthalic anhydride

Materials

2,3-Dimethylbuta-1,3-diene (Freshly distilled)	2.05 g
Maleic anhydride	2.45 g

Procedure

A mixture of 2,3-dimethylbuta-1, 3-diene (2.05 g, 0.025 mol) and finely powdered maleic anhydride (2.45 g, 0.025 mol) is allowed to react. The reaction occurs in few minutes as indicated by evolution of heat. Keep the reaction mixture at room temperature for 15-20 min and extract with cold water to remove excess of maleic anhydride. The residual product is dried and recystallised from petroleum ether, m.p. 78-79°. Yield is almost quantitative.

Notes

1. The Diels-Alder reaction of 2,3-dimethylbuta-1,3-diene with naphthoquinone (5 hr. refluxing in ethanol) gives 91% yield of cis-1,4,4a,9a-tetrahydro-2,3-dimethyl-9,10-anthraquinone.

$$\text{Naphthoquinone} + H_2C{=}C(CH_3){-}C(CH_3){=}CH_2 \xrightarrow[\text{5 hr.}]{\Delta,\ \text{EtOH}} \text{product}$$

3.24 MISCELLANEOUS APPLICATIONS OF SOLID STATE REACTIONS

Oxidation of Alcohols to Carbonyl Compounds

A number of methods are available to accomplish the oxidation of alcohols to carbonyl compounds. The conventional oxidizing reagents normally employed include peracids, peroxides, MnO_2, $KMnO_4$, CrO_3, K_2CrO_7 and $K_2Cr_2O_7$. Following are given some of the convenient procedures for the oxidation of alcohols to carbonyl compounds.

(i) Selective and solventless oxidation of alcohols to carbonyl compounds using montmorillonite K 10 clay-supported iron (III) nitrate under solvent free conditions using MW irradiation (R.S. Varma and R. Dahiya, Tetrahedron Lett., 1997, **38**, 2043) has been developed.

$$\begin{matrix} R_1 \\ R_2 \end{matrix} \rangle CH - OH \xrightarrow[\text{MW, 30-60 sec}]{\text{Clayfen}} \begin{matrix} R_1 \\ R_2 \end{matrix} \rangle C = O$$

R_1	R_2	Time(sec.)	Yields(%)
Ph	H	15	92
Ph	Et	30	87
4-MeC_6H_4	H	15	94
4-$MeOC_6H_4$	H	10	96

(ii) On oxidation with manganese dioxide-silica, using MW irradiation, alcohols are oxidised to carbonyl compounds (R.S. Varma, Clean Products and processes, 1999, 139 and the references cited there in)

$$\begin{matrix} R_1 \\ R_2 \end{matrix} \rangle CH - OH \xrightarrow[\text{MW, 20-50 sec}]{MnO_2\text{-silica}} \begin{matrix} R_1 \\ R_2 \end{matrix} \rangle C = O$$

R_1	R_2	Time(sec.)	Yields(%)
Ph	H	20	88
Ph	Et	60	84
4-MeC_6H_5	H	45	81
4-$MeOC_6H_5$	H	30	83
Ph	Ph	45	85

(iii) On oxidation with chromium trioxide impregnated on wet alumina oxidises alcohols to carbonyl compounds. The oxidation is conducted by irradiation with microwaves for 30-40 seconds). The mole ratio of the substrate and the reagent used is 1:2 (R.S. Varma and R.K. Saini, Tetrahedron Lett., 1998, **39**, 1481)

$$R_1R_2CH-OH \xrightarrow[\text{MW, 30-40 Sec.}]{\text{Wet } CrO_3-Al_2O_3} R_1R_2C=O$$

R_1	R_2	Time(sec.)	Yields(%)
Ph	H	—	76
Ph	Ph	35	87
Ph	Me	—	84
4-MeC_6H_4	H	—	83
4-$NO_2C_6H_4$	H	40	76

(iv) Alcohols can also be oxidised to carbonyl compounds by using iodobenzene diacetate (IBD) as an oxidant under solvent free conditions using MW irradiation in quantitative yield (R.S. Varma, R.Dahiya and R.K. Saini, Tetrahedron Lett, 1998, 38, 7029)

$$R_1R_2CH-OH \xrightarrow[\text{MW, 0.5-2 min}]{\text{IBD/Alumina}} R_1R_2C=O$$

R_1	R_2	Time(min.)	Yields(%)
Ph	H	1.0	94
Ph	Et	2.0	89
4-MeC_6H_4	H	1.0	92
4-MeC_6H_4	H	2.0	95

IBD is commercially available. It can also be prepared by treatment of iodobenzene with peroxyacetic acid with 30% H_2O_2 and acetic anhydride at 40° (J.G. Sharefkin and H. Haltziaman, org. synth, collective volume 1973, **5**, 658, 660)

Oxidation of Sulfides to Sulfoxides

Sulfides on oxidation with sodium periodate give selectively sulfoxides using 10% sodium periodate 'doped' on silica gel and subjecting to microwave irradiation (R.S. Varma, R.K. Sani and H.M. Meshram, Tetrahedron Lett., 1997, **38**, 6525)

$$R-S-R_1 \xrightarrow[\text{Microwave 1-3 min}]{10\%\ NaIO_4\text{-silica(1.7 eq)}} R-\overset{O}{\overset{\|}{S}}-R_1$$

72-93%

R	R_2	Time(min)	Yields(%)
n-Bu	nBu	1.0	72
Ph	Me	2.5	82
Ph	Ph	2.4	93
$PhCH_2$	Me	2.5	87
$PhCH_2$	$PhCH_2$	1.0	72

The oxidation of sulfides to sulfoxides can also be effected by iodobenzene diacetate (IBD) supported on alumina as an oxidant under solvent free conditions using MW irradiation (R.S. Varma, R.K. Saini and R. Dahiya, J.Chem. Soc.(S), 1998, 120)

$$R_1-S-R_2 \xrightarrow[\text{Micro wave 40-90 sec}]{PhI(OAc)_2\text{-Alumina}} R-\overset{\overset{\displaystyle O}{||}}{S}-R_1 \quad 80\text{-}90\%$$

R_1	R_2	Time(sec)	Yields(%)
t-Pr	i-Pr	40	80
n-Bu	n-Pr	40	82
Ph	Me	45	82
Ph	Ph	90	88
$PhCH_2$	Ph	90	86
$PhCH_2$	$PhCH_2$	90	90

IBD is commercially available. It can also be prepared as described in the oxidation of alcohols to carbonyl compounds (See Page 3.25)

Sulfides (alkyl, aryl and cyclic) can be rapidly oxidised to the corresponding sulfoxides in good yield upon MW irratidiation with Fe(III) nitrate impregnated on clay (clayfen) and solvent free conditions. The conversion can also be effected in refluxing methylene chloride but requires much longer time (R.S. Varma and R. Dahiya, Synth. Commun., 1998, **28**, 4087)

$$R_1-S-R_2 \xrightarrow[\text{Micro wave 0.5-1 min}]{\text{Clayfen}} R_1-\overset{\overset{\displaystyle O}{||}}{S}-R_2 \quad 80\text{-}90\%$$

R_1	R_2	Time(min)	Yields(%)
i-Pr	i-Pr	0.5	89
n-Bu	n-Bu	—	86
Ph	Me	0.25	91
Ph	Ph	1.0	75
Ph	$PhCH_2$	1.0	85
$PhCH_2$	$PhCH_2$	1.0	15

Pinacol-Pinacolene Rearrangement

This rearrangement is effective by microwave irradiation of gem-diols with Al^{3+} -montmorillonite K clay for 15 min. (E-Gutierrez, A. Loupy, G. Bram and E. Ruiz-Hitzky, Tetrahedron Lett., 1989, **30**, 945)

$$\text{gem-Diols (Pinacols)} \xrightarrow[\text{MW, 15 min}]{Al^{3+}\text{–Montmorillonite}} \text{Pinacolones } 98\text{-}99\%$$

Beckmann Rearrangement

The Ketoximes on heating in microwave owen with Montmorillonite K Clay give the rearranged product (A.I. Bosch, P. de lacruez, E. Dilz-Barra, A. Lupy and F.Lanya, Synlett., 1995, 1259).

$$\underset{\text{Ketoximes}}{RR'C{=}NOH} \xrightarrow[\text{MW, 7-10 min}]{\text{Montmorillonite k Clay}} \underset{\text{Analides}}{R-\overset{O}{\overset{\|}{C}}-NH-R'}$$

Crossed Cannizzaro Reaction

The crossed cannizzaro reaction between an aldehyde and formalin can be performed by heating in a MW oven with Ba $(OH)_2$. $8H_2O$ (R.S. Varma, K.P. Naicker and P.J. Liesen, Tetrahedron Lett., 1998, **39**, 8439)

$$RCHO + (CH_2O)_n \xrightarrow[\text{Microwave, 5 min}]{Ba(OH)_8 8H_2O} \underset{80\text{-}90\%}{RCH_2OH} + \underset{1\text{-}20\%}{RCOOH}$$

See also preparation of benzyl alcohol (Section 3.6).

Benzimidazoles

These are easily prepared by the condensation of ortho esters with o-phenylenediamines in presence of KSF Clay using focused microwave irradiation (D. Villemin, 17. Hammadi and B. Martin, Synth. Commun., 1996, **26**, 2895)

$$(EtO)_3C + \text{o-}(H_2N)_2C_6H_4 \xrightarrow[\text{MW, 2-3 min}]{\text{KSF Clay}} \text{benzimidazole (N, N–H)}$$

Alkylation of Reactive Methylene Compounds

In this procedure the compound containing reactive methylene group is heated in a microwave oven with alkyl halide in presence KOH, K_2CO_3 and tetra butyl ammonium chloride as a PTC (D. Runhua, W. Yuliag and J. Yaohong, Synth. Commun., 1994, **24**, 111)

$$\underset{\text{Ethylacetoacetaic}}{CH_3COCH_2CO_2Et} \xrightarrow[\text{Microwave, 3 min}]{RX,\ KOH,\ K_2CO_3,\ PTC} CH_3COCH(R)-CO_2Et$$

$$R = CH_3CH_2CH_2,\ PhCH_2,\ m-CH_3OC_6H_4CH_2,\ p\text{-}ClC_6H_4CH_2CH_2(CH_2)_5$$

$$\underset{\text{Ethyl phenyl mercaptoacetate}}{C_6H_5\ SCH_2CO_2Et} \xrightarrow[\text{Microwave, 3 min}]{RX \cdot KOH-K_2CO_3,\ PTC} C_6H_5\ SCH(R)-CO_2Et$$

$$R = CH_3CH_2 \quad CH_2,\ p\text{-}ClC_6H_4CH_2\text{–},\ m\text{-}CH_3OC_6H_4CH_2 -$$

Aziridines

A. Saoudi, J. Hamelin and H. Benhaoua, J. Chem. Res. (S), 1996, 492).

$$X-CH(Br)-CH(Br)-R \xrightarrow[\text{Bentonite MW, 3-5 min}]{R'NH_2} \text{aziridine (X, H / R, H; N–R')} + \text{aziridine (H, X / R, H; N–R')} + \begin{matrix} RCH=NR' \\ RCH=CXBr \end{matrix}$$

X = electron withdraing group

For more examples of microwave assisted reaction in solid state see V.K. Ahluwalia and R.S. Verma in Alternative Energy Processes in Chemical Synthesis, Chapter-4. Narosa Publishing House, 2028.

3.25 CONCLUSION

Solid states organic transformations, as the name implies do not require any solvent and are considered to be the best examples of green transformations. Some of the reactions which can be conducted in solid state include Claisen rearrangement, Dieckmann condensation, Paal-Knorr synthesis, oxidation of alcohols to carbonyl compounds, sulfides to sulfoxides, Baeyer-villiger oxidation, Diels-Alder reaction and Beckmann rearrangement. Besides, this procedure has also been used for a large number of organic transformations like conversion of aldehydes into nitriles. Cyanides into amides, benzoin into benzil, benzaldehyde into benzyl alcohol, aldehydes into phenols and reduction of carbonyl group

Chapter 4

Photochemical Transformations

4.1 CIS-AZOBENZENE

The azobenzene as ordinarily obtained is the trans-isomer. It can be isomerised photochemically into cis form by irradiation with a fluorescent lamp or even by sunlight.

$$H_5C_6-\ddot{N}=\ddot{N}-C_6H_5 \xrightarrow{h\nu} \underset{H_5C_6}{}\ddot{N}=\ddot{N}\underset{C_6H_5}{}$$

Trans-azobenzene Cis-azobenzene

Materials

Azobenzene (trans)	1 g
Petroleum ether (b.p. 40-60°)	50 mL

Procedure

Azobenzene (trans) (1 g) is dissolved in petroleum ether (50 mL) in a beaker (250 mL capacity). The flask is placed in bright sunlight. The reaction takes 4 days for completion (1 hr in uv light by supporting a Honovia fluorescent lamp above the surface of the liquid in the beaker). In the meanwhile, a chromatographic column (approx 20 cm × 1.8 cm) from activated alumina (grade I, 50 g) is prepared by making a slurry in petroleum ether and pouring into the column. A circular wall fitting filter paper is placed at the top of the column. The irradiated solution is poured on the filter paper in the column with the help of a glass rod. Precaution is taken so that the column is not disturbed. The column is eluted with petroleum ether (b.p. 40-50°, 100 mL). A sharp coloured band of cis-azobenzene (about 2 cm) appears at the top of the column while a diffuse coloured region of trans azobenzene moves down the column. The upper band is protected by covering with a cabon paper. This will prevent the reconversion of cis into trans form. The column is extruded and the upper band is cut and shaken with petroleum either (150 mL) containing methanol (1.5 mL). The solution is filtered washed with water (2 × 15 mL), dried (Na_2SO_4) and solvent removed. The residual coloured product is pure cis-azobenzene, m.p. 71.5°.

Note

1. The puriy of cis azobenzene can be confirmed by recording the uv absorption spectrum in ethanol solution as soon as possible after its isolation. Cis-azobenzene has λ_{max} 281 nm (ε 5260); trans azo-benzene has λ_{max} 320 nm (ε 21300) in ethanol solution.
2. The azobenzene (trans) required is prepared as follows:

   To a suspension of magnesium turnings (1 g), nitrobenzene (1.8 mL), methanol (35 mL) is added a small crystal of iodine in a round bottomed flask. The flask is fitted with a reflux condenser. If the reaction does not commence in 2-3 minutes, warm the mixture (water-bath) to start the reaction. In case the reaction becomes too vigorous, use a cold water bath for few seconds. When most the magnesium has reacted, more magnesium turnings (1 g) is added. Let the reaction proceed (as above). Finally the reaction mixture is heated on a water–bath cooled and poured into water (65 mL). The mixture is acidified with glacial acetic acid (until the mixture is acidic to congo red). Cool the mixture (ice-bath) filter the separated azobenzene and crytallise from ethanol (90%). Yield 1 g (34%), m.p. 67-68°.

4.2 BENZOPINACOL

It is obtained by photoreduction of benzophonone by irradiation in propan-2-ol. Sunlight or a medium pressure mercury are lamp is used. The mechanism involved is given below. The propan-2-ol acts as a hydrogen donor and itself is oxidisd to acetone.

$$\underset{\text{Benzophenone}}{(C_6H_5)_2C{=}O} \xrightarrow{h\nu} \underset{\text{(singlet)}}{(C_6H_5)_2C{=}\dot{O}} \longrightarrow \underset{\text{(triplet)}}{(C_6H_5)_2\dot{C}{=}\dot{O}}$$

$$\underset{\text{(triplet)}}{Ph_2\dot{C}-\dot{O}} + H-\underset{CH_3}{\overset{CH_3}{\dot{C}}} \longrightarrow Ph_2\dot{C}-OH + \cdot\underset{CH_3}{\overset{CH_3}{C}}-OH$$

$$\underset{\text{(triplet)}}{Ph_2\dot{C}-\dot{O}} + HO-\underset{CH_3}{\overset{CH_3}{C}}\cdot \longrightarrow Ph_2\dot{C}-OH + O{=}C(CH_3)_2$$

$$2Ph_2\dot{C}-OH \longrightarrow Ph-\underset{Ph}{\overset{OH}{C}}-\underset{Ph}{\overset{OH}{C}}-Ph$$

Materials

Benzophanone	2.5 g
Isopropyl alcohol	10 mL

Procedure

Benzophenone (2.5 g) is dissolved in isopropyl alcohol (10 mL) by warming followed by addition of one drop of acetic acid. More isopropyl alcohol is added till its level is about 1 cm below the flask Joint (R.B of 25 mL capacity is used). The flask is stopperd (rubber stopper) and placed in direct sunlight. Within 20 hrs. colourless crystals of benzopinacol start separating. The flask is allowed to remain in sunlight until no further solid appears to separate (8 –10 days). The solution is cooled (ice-water), separated product filtered and crystalloid from glacial acetic acid yield 2.6 g (98 %), M.P. 185-86°.

Notes

1. The reaction taken much shorter time (3-4 hr) if medium pressure mercury are lamp is used.
2. Benzopinacol is also obtained by reductive dimerisation of benzopinacol. The method involves refluxing a mixture of benzophenone, zinc dust, glacial acetic acid and water for 2 hrs. followed by cooling. Yield in only 35%.

4.3 MALEIC ACID

It is the cis-(or z) isomer and is thermodynamically less stable than fumanic acid (which is trans-or E isomer). It is obtained the photochemical isomensation of trans isomer (fumaric acid)

$$\underset{\text{(trans or E-isomer)}}{\underset{\text{Fumaric acid}}{\text{HOOC(H)C}=\text{C(H)COOH}}} \xrightarrow{h\nu} \underset{\text{(cis-or z isomer)}}{\underset{\text{Malic acid}}{\text{HOOC(H)C}=\text{C(H)COOH}}}$$

The procedure for the conversion in similar to that used for the conversion of trans azobenzene to cis-azobenzene (Section 4.1)

Notes

1. Fumanic acid, m.p. 287° is obtained by the debromination of bromosuccinic acid with alcoholic potash or by the condensation of glyoxylic acid with malonic acid (**Knoevanagel reaction**), followed by decarboxylation.

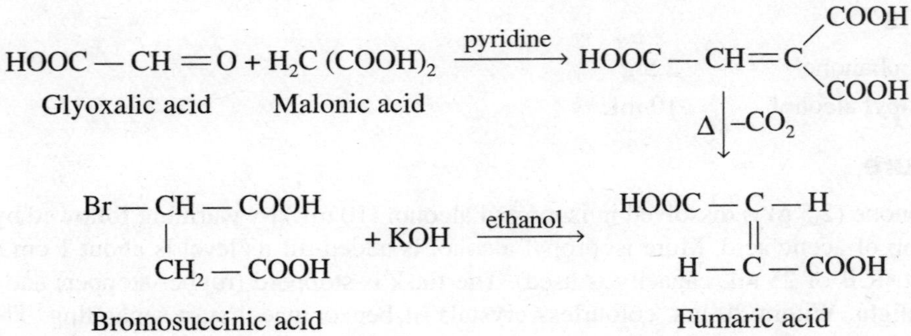

$$HOOC-CH{=}O + H_2C(COOH)_2 \xrightarrow{\text{pyridine}} HOOC-CH{=}C(COOH)_2$$

Glyoxalic acid Malonic acid

$$\downarrow \Delta, -CO_2$$

$$Br-CH(COOH)-CH_2-COOH + KOH \xrightarrow{\text{ethanol}} HOOC-C-H \,{=}\, H-C-COOH$$

Bromosuccinic acid Fumaric acid

2. Maleic acid, m.p. 138-39° is also obtained by heating malic acid at 250°, by the following sequence of reaction

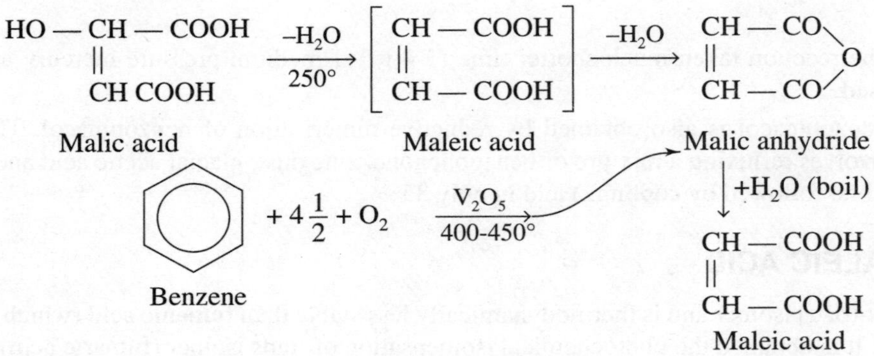

$$HO-CH-COOH \,/\!/\, CH\,COOH \xrightarrow[250°]{-H_2O} [CH-COOH \,{=}\, CH-COOH] \xrightarrow{-H_2O} CH-CO \,{=}\, CH-CO \;(>O)$$

Malic acid Maleic acid Malic anhydride

$$\text{Benzene} + 4\frac{1}{2} + O_2 \xrightarrow[400\text{-}450°]{V_2O_5} \text{Malic anhydride} \xrightarrow{+H_2O\ (\text{boil})} CH-COOH \,{=}\, CH-COOH$$

Benzene Maleic acid

It is also obtained industrially by the oxidation of benzene with air at 400-450° in presence of V_2O_5 followed by hydrolysis of the malic anhydride.

4.4 METHYL α-NAPHTHYL ACETATE

It is prepared by the photochemical reaction of α-naphthoyl chloride with diazomethane in presence of methanol.

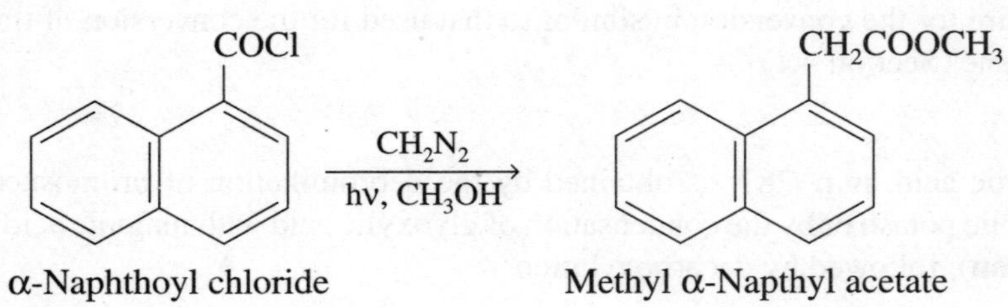

$$C_{10}H_7COCl \xrightarrow[h\nu,\ CH_3OH]{CH_2N_2} C_{10}H_7CH_2COOCH_3$$

α-Naphthoyl chloride Methyl α-Napthyl acetate

This reaction is **Arndt-Eistert synthesis**[1], which is a general method for converting an carboxylic acid into its next higher homologue via the formation of acid chloride

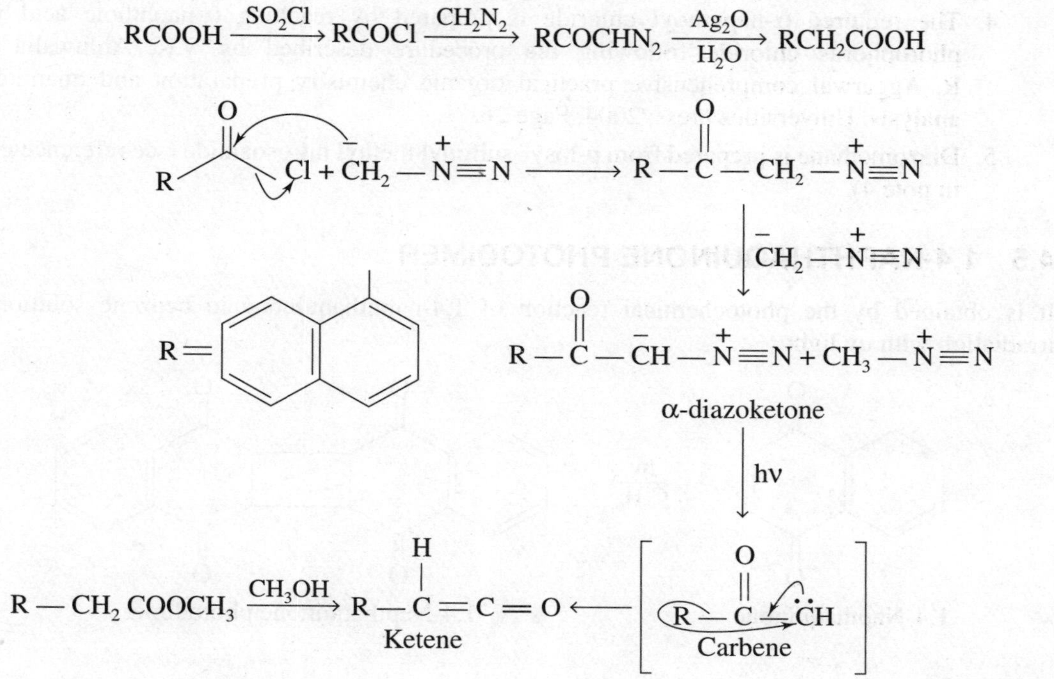

$$\mathrm{RCOOH} \xrightarrow{SO_2Cl} \mathrm{RCOCl} \xrightarrow{CH_2N_2} \mathrm{RCOCHN_2} \xrightarrow[H_2O]{Ag_2O} \mathrm{RCH_2COOH}$$

$$\mathrm{R{-}C(=O){-}\overset{-}{Cl}} + \mathrm{CH_2{-}\overset{+}{N}{\equiv}N} \longrightarrow \mathrm{R{-}C(=O){-}CH_2{-}\overset{+}{N}{\equiv}N}$$

$$\xrightarrow{\overset{-}{C}H_2{-}\overset{+}{N}{\equiv}N} \mathrm{R{-}C(=O){-}\overset{-}{C}H{-}\overset{+}{N}{\equiv}N} + \mathrm{CH_3{-}\overset{+}{N}{\equiv}N}$$

R = (α-naphthyl)

α-diazoketone

$$\xrightarrow{h\nu} \left[\mathrm{R{-}C(=O){-}\ddot{C}H}\right] \text{ Carbene} \longrightarrow \mathrm{R{-}C(H){=}C{=}O} \text{ Ketene} \xrightarrow{CH_3OH} \mathrm{R{-}CH_2\,COOCH_3}$$

Materials

α-Naphthoyl chloride	5 g
$CH_2\,N_2$ in methanol	......
$CH_3\,OH$	20 mL

Procedure[2]

α-Napthoyl chloride (5 g) is dissolved in methanol (20 mL) and reaction mixture is placed in sunlight for 5 days for completion (3 hr. with uv lamp irradiation). The methanol is removed (vacuo) to give methyl α-naphthyl acetate as an oily product yield 5 g.

Notes

1. F. Arndt and B. Eistert, Ber, 1935, 68, 200
2. The procedure followed is similar to that described by A.B. Smith, Chem. Commun. 1974, 695
3. α-Naphthoyl acetate is characterised by saponification to give α-Naphthyl acetic acid, m.p. 130°

4. The required α-naphthoyl chloride is prepared by reaching α-naphthoic acid with phorophorus chloride following the procedure described by V.K. Ahluwalia and R. Aggarwal comprehensive practical organic chemistry preparation and quantitative analysis, Universities press, 2004, Page 26.
5. Diazomethane is prepared from p-tosyl-sulfonyl methyl nitrosoamide (see reference given in note 4).

4.5 1,4-NAPHTHAQUINONE PHOTODIMER

It is obtained by the photochemical reaction of 1,4-naphthaquinone in benzene solution by irradiation with uv light.

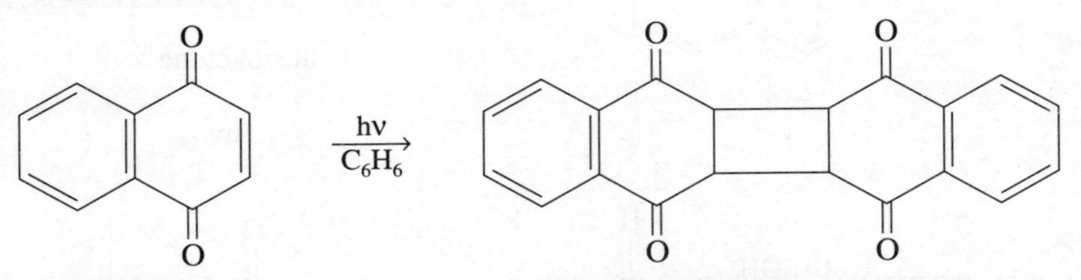

Materials

1, 4-Naphthaquinone	3 g
Benzene (thiophene free)	50 mL

Procedure

Take a solution of 1,4-naphthaquinone (3 g) in thiophene free benzene (50 mL) in a R.B flask (50 mL capacity). Pass a slow stream of nitrogen through the solution to remove any air or oxygen. Stopper the flask and place in sun light. Wrap the rubber stopper with aluminium foil. After 4-5 days, a solid product separates. Filter the separated product and expose the filterable to sunlight after passing nitrogen gas (as earlier). Collect the separated product after 4-5 days. Repeat the process 4-5 times (total of approximately 30 days exposure). Total yield 0.44g (13%). M.P. 244-246°.

Note

1. The photodimensation can also be done by using a 100 watt medium pressure mercury arc lamp for 6 hrs. the yield is about 12%.

4.6 9-PHENYLPHENANTHRENE

It is obtained by irradiation of triphenyl ethylene in cyclohexane in presence of iodine using a photochemical reactor

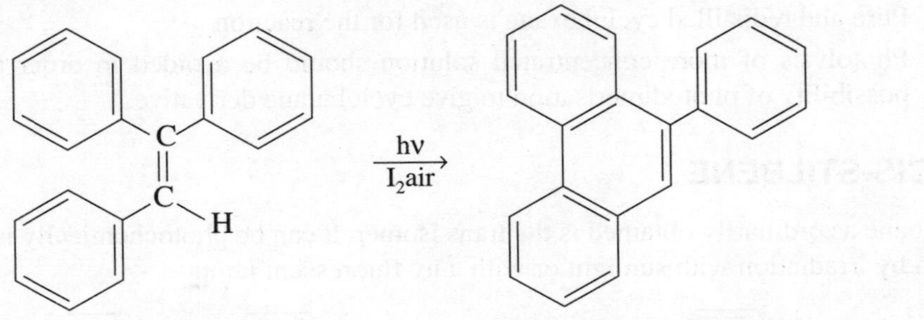

$$\text{Triphenyl ethylene} \xrightarrow[I_2 air]{h\nu} \text{9-Phenyl phenanthrene}$$

Triphenyl ethylene

9-Phenyl phenanthrene

Materials

Triphenylethylene	2.56 g
Iodine	0.127 g
Cydohexane	100 mL

Procedure

A stirred solution of triphenylethylene (2.56 g, 0.01 mol) and iodine (0.127 g, 0.0005 mol) in cyclohexane (100 mL) is irradiated in a photochemical reactor using 100-W medium-pressure mercury are lamp. The reaction mixture is exposed to the atmosphere and irradiation continued for about 24 hrs. The solution is evaporated (vacuum) and the residual product purified by column chromatography using neutral alumina (16-18 g) as adsorbent and elution with cydohexane. The product is crystallised from ethanol yield 2.16 g (85%). M.P. 104-105°.

Notes

1. The required triphenylethylene is obtained by the reaction of benzophenone with benzyl magnesium chloride under anhydrous conditions followed by treatment of the adduct with dil $H_2\ SO_4$

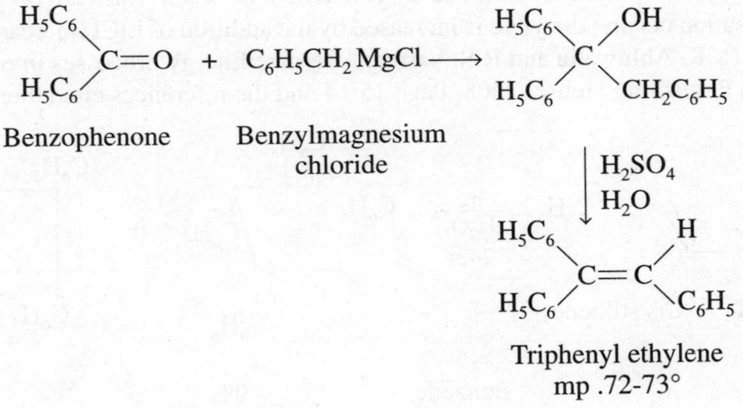

$$(H_5C_6)_2C{=}O + C_6H_5CH_2MgCl \longrightarrow (H_5C_6)_2C(OH)CH_2C_6H_5 \xrightarrow[H_2O]{H_2SO_4} (H_5C_6)_2C{=}C(H)C_6H_5$$

Benzophenone

Benzylmagnesium chloride

Triphenyl ethylene
mp .72-73°

2. Pure and redistilled cyclohexane is used for the reaction.
3. Photolysis of more concentrated solution should be avoided in order to minimise the possibility of photodimerisation to give cyclobutane derivative.

4.7 CIS-STILBENE

The stilbene as ordinarily obtained is the trans isomer. It can be photochemically isomerised to the cis form by irradiation with sunlight or with a uv fluorescent lamp.

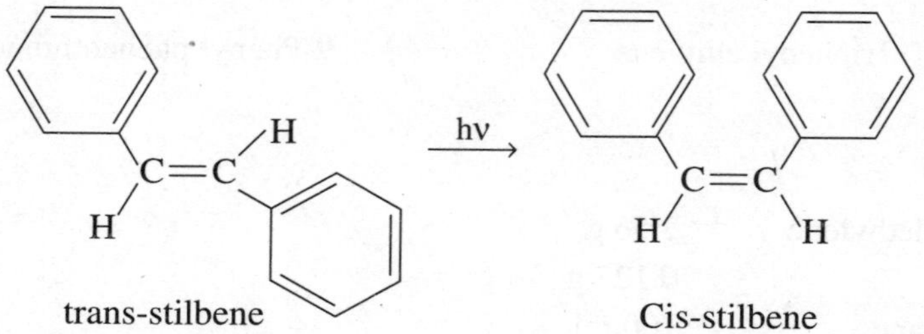

$$\text{trans-stilbene} \xrightarrow{h\nu} \text{Cis-stilbene}$$

trans-stilbene   Cis-stilbene

The procedure for the conversion is similar to that used for the conversion of trans azobenzene to cis azobenzene (Section 4.1) In this case the cis form is obtainer in 90%; the trans is 10%.

Notes

1. The trans-stilbene, m.p 124° is prepared by the **Clemmensen reduction** of benzoin (Shriner, Berger, Org. Synth. coll. Vol. III, 1955, 786) or by the Witting reaction of benzaldehyde with benzyltriphenyl phosphorium chloride(see Section 2.4.1) (in this case a mixture of cis and trans form is obtained)
2. The cis-form obtained by the above procedure solidifies at-5°, b.p. 96°/1 atm, uv (max) (99% ethanol) 278 nm (ε 10, 200)
3. A mixture of cis and trans form is obtained by the decarboxylation of phenylcinnamic acid (Wheelor, Battle de Patron, J. Org. Chem. 1965, 30 1473).
4. In case isomerisation of stilbene is conducted in water (instead of petroleum ether) dimerisation occurs; the yield is increased by the addition of LiCl (increasing hydrophobic effect) (V.K. Ahluwalia and R.S. Varma, Alternate Energy processes in organic synthesis, Narosa Publishing House, 2008, Page 15.14 and the references cited there in)

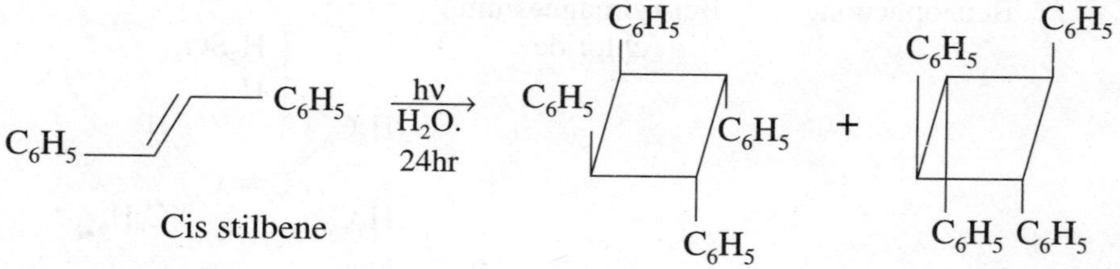

$$C_6H_5\text{-}CH{=}CH\text{-}C_6H_5 \xrightarrow[24hr]{h\nu,\ H_2O} \text{cyclobutane dimer } (C_6H_5)_4C_4H_4 + \text{cyclobutane dimer } (C_6H_5)_4C_4H_4$$

Cis stilbene

Benzene	0%	0%
Water	12%	10%
Water-LiCl	25%	17%

4.8 MISCELLANEOUS APPLICATIONS OF PHOTOLYSIS

1. **Photochemical cycloaddition reactions**

(a) [2 + 2] Cycloadditions

$$\begin{matrix} H_2C = CH_2 \\ H_2C = CH_2 \end{matrix} \xrightarrow{h\nu} \begin{matrix} CH_2 - CH_2 \\ | \quad\quad | \\ CH_2 - CH_2 \end{matrix}$$

Ethylene Cyclobutane

(b) [4 + 2] Cydoadditions

$$\text{Butadiene} + \text{Ethylene} \xrightarrow{h\nu} \text{Cyclohexene}$$

Butadiene Ethylene Cyclohexene

(c) Intramolecular photocyclisation

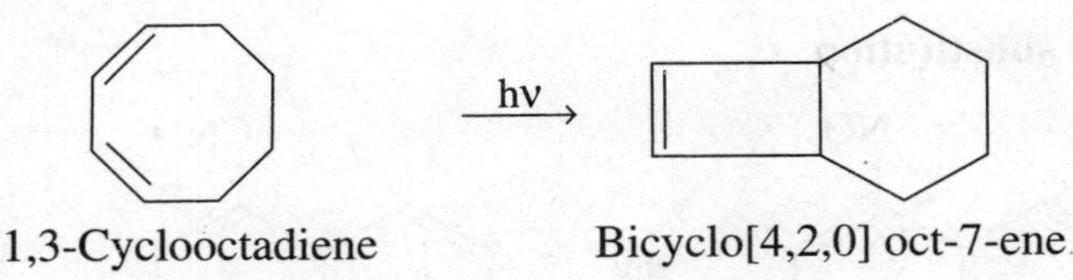

1,3-Cyclooctadiene Bicyclo[4,2,0] oct-7-ene

2. **Paterno-Buchi reaction**

It involves the photochemical cycloaddition of carbonyl compounds to olefins to give oxetanes (oxa-cyclobutanes)

$$\underset{\text{Ketone}}{R - \overset{\overset{O}{||}}{C} - R'} + \underset{\text{2-methyl butene2}}{(H_3C)(H)C = C(CH_3)_2} \xrightarrow{h\nu} \begin{matrix} & & H & \\ & & | & \\ & O - & C - & CH_3 \\ & | & | & \\ R - & C - & C - & CH_3 \\ & | & | & \\ & R' & CH_3 & \end{matrix} + \begin{matrix} & & CH_3 & \\ & & | & \\ & O - & C - & CH_3 \\ & | & | & \\ R - & C - & C - & H \\ & | & | & \\ & R' & CH_3 & \end{matrix}$$

3. **Photochemical cycloadditions in water**

(a) Stilbene photodimerise in water in presence of LiCl (M.S. Sayamala and V. Ramamurthy, J.Org. Chem., 1986, **51**, 3712)

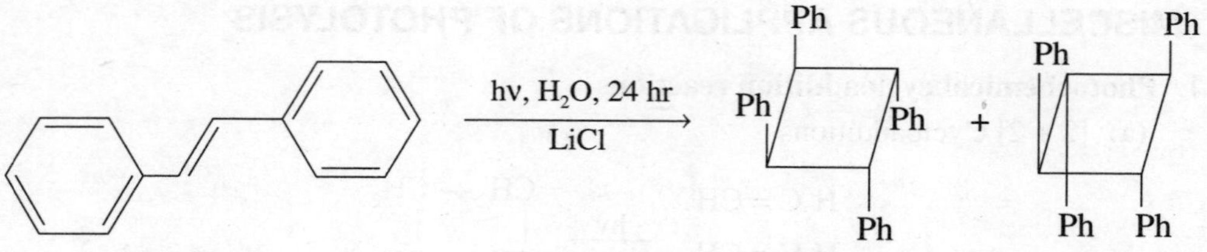

hν, H_2O, 24 hr, LiCl

Stilbene

Ph Ph Ph Ph + Ph Ph Ph Ph

25% 17%

(b)

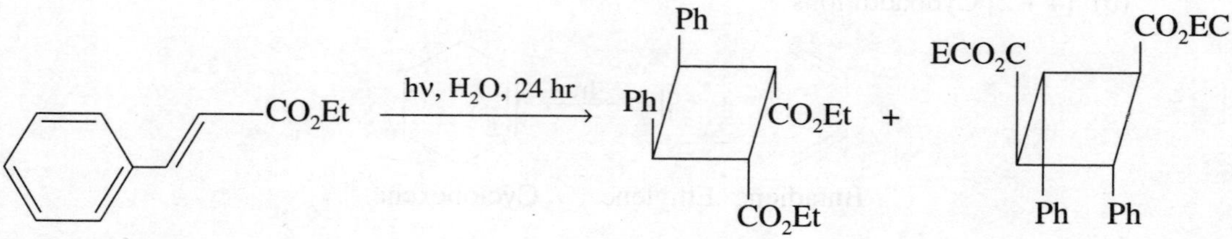

hν, H_2O, 24 hr

CO_2Et

Ethylcinnamate

Ph Ph CO_2Et CO_2Et + ECO_2C CO_2EC Ph Ph

4. **Phoinduced substitution**

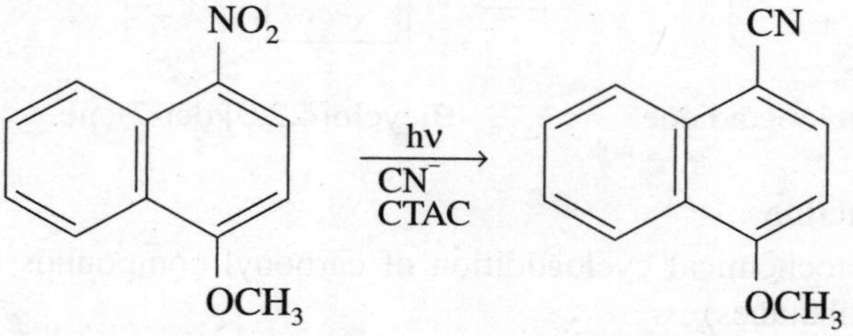

NO_2 OCH_3 —hν, CN^-, CTAC→ CN OCH_3

4-Methoxy-1-nitro napthalene 4-Methoxy-1-napthalene carbonitrile

[CTAC = hexadecyl (trimethyl) ammoniun bromide]

5. **Photorearrangement**

(K. Yamada, K. Shigehiro, T. Kujozuka and H. Lida, Bull. Chem. Soc. Jpn., 1978, **51**, 2447)

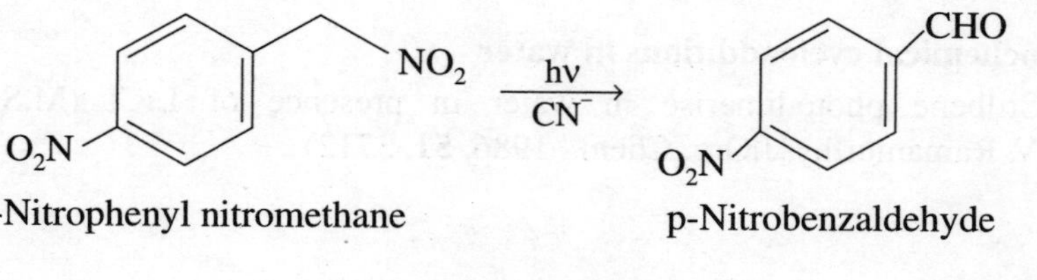

O_2N NO_2 —hν, CN^-→ O_2N CHO

p-Nitrophenyl nitromethane p-Nitrobenzaldehyde

6. **Photochemical reaction in solid state**

   A tycical example is photo irradiation of cinnamic acid (single crystal) to give truxillic acid (V. Ramamurty, Ed., Photochemisty of organized and constraint media, VCH, Weinheimn, Germany, 1991)

HOOC COOH $\xrightarrow[\text{Solid state}]{h\nu}$ HOOC CHOO

Cinnamic acid Truxillic acid

7. **Photochlorination**

   (F. Broich, Fette and Seifen, Anstrichm, 1970, **17**, 22)

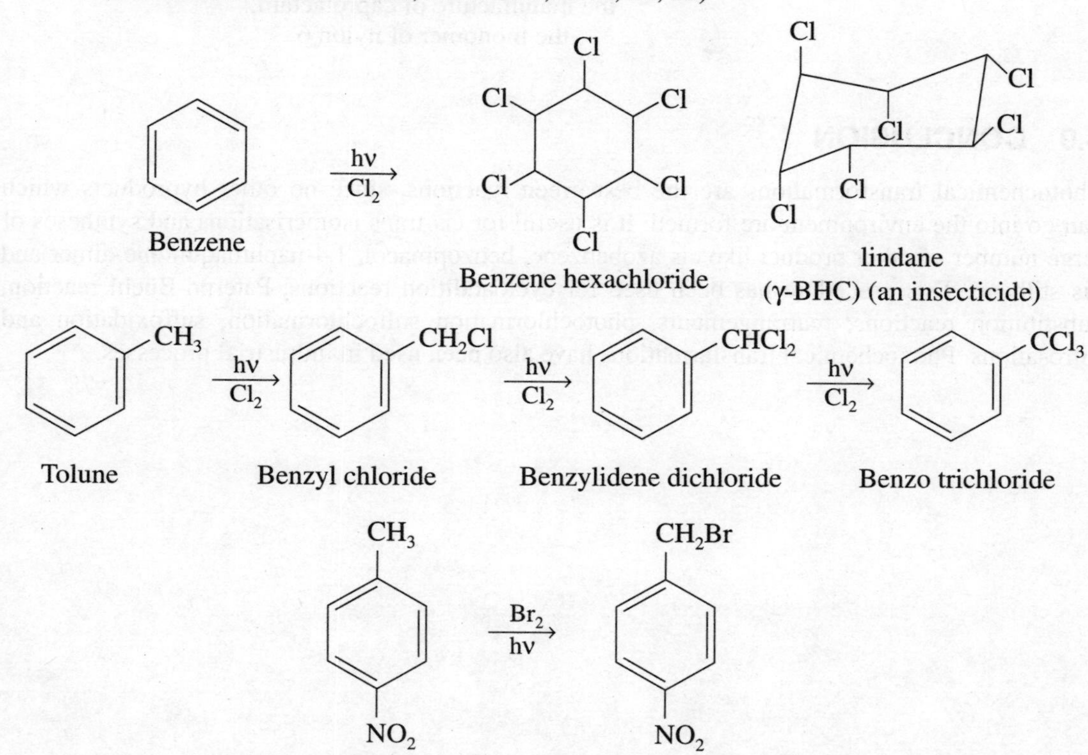

8. **Photo Sulfochlorination**

   (K. Lindner, Tenside, Textihilfsmittel, Waschrohstoffe, Wissenschafil, Verlagsges, Stuttgart, 1964, **Vol. 1**, P. 705)

$$RH + SO_2Cl_2 \xrightarrow{h\nu} RSO_2Cl + HCL$$

   RH = mixture of n-alkenes

9. **Photochemical Sulfoxidation**

   (L. Orthner, Angew. Chem. 1950, **62**, 302; R. Grat, Justus Liebigs Ann. Chem., 1952, **50**, 578)

$$RH + SO_2 + 1/2\ O_2 \xrightarrow{h\nu} RSO_3H$$

$$RH = C_{14\text{-}18}$$

10. **Photonitrosation**

    (P. Turner, Int. Chins, 1970, **9(5,6)**, 51; P. Hurine, P.E. Turner, Chem. Process Eng. 1967 **(11)**, 96)

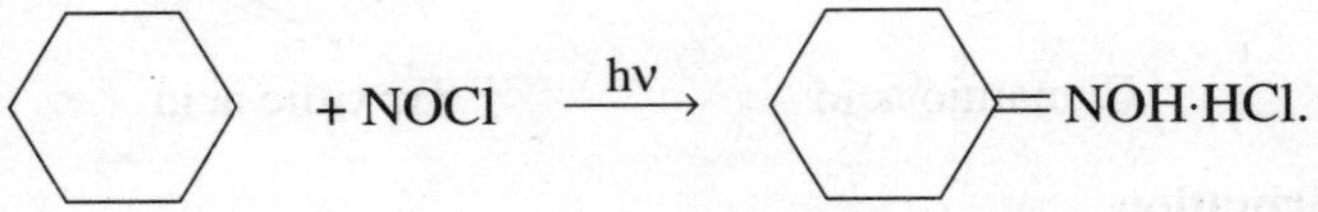

Cyclohexane + NOCl $\xrightarrow{h\nu}$ cyclohexane=NOH·HCl.

Cyclohexane

Cyclohexane oxime used for the manufacture of caprolactam, the monomer of nylon 6

4.9 CONCLUSION

Photochemical transformations are the best green reactions, since no other byproducts which can go into the environment are formed. It is useful for cis-trans isomerisations and syntheses of large number of other product like cis azobenzene, benzopinacol, 1,4-naphthaquinone dimer and cis stilbene. This procedure has been used for cycloaddition reactions, Paterno-Büchi reaction, substitution reactions, rearrangements, photochlorination sulfochlorination, sulfoxidation and nitrosations. Phoxochemical transfarmations have also been used in indrustrial processes.

Chapter 5

Transformations Using Phase Transfer Catalysts

5.1 BENZOIC ACID

It is normally obtained by the oxidation of toluene by alkaline permanganate solution. However even after refluxing for longer time, the yield is only 40-50%. This is because toluene is not soluble in water and so the $KMnO_4$ does not work efficiently. However, oxidation with $KMnO_4$ in solvents like acetic acid 1-butonol or pyridine give better results. It has been found that oxidation with solution of permanganate can be effected in presence of phase transfer catalysts (e.g. tetraalkyl ammonium or phosphonium salts). The permanganate anion strongly associates with quaternary cations and is readily extracted into organic solvents by the reagents (W.A. Gibson and R.A. White Anal. Chem. Acta, 1955, **12**, 413; H.W. Herriott and D. Picker, Tetrahedron Lett. 1974, 1511). A very small amount of PTC is sufficient to transfer all permanganate from aqueous to organic phase and oxidise toluene to benzoic acid in good yields.

$$\text{Toluene } (C_6H_5CH_3) \xrightarrow[\text{PTC}]{(O)\ \text{alk. } KMnO_4} \text{Benzoic acid } (C_6H_5COOH)\ 80\text{-}90\%$$

Materials

Toluene	2.5 mL
Potassium permanganate	4 g
Sodium carbonate (2N)	1.6 mL
Cetyltrimethyl ammonium chloride	0.1 g

Procedure

Potassium permanganate solution (4 g. dissolved in water) is added to a refluxing mixture of toluene (2.5 mL), sodium carbonate solution (2N, 1.6 mL) and cetyltrimethyl ammonium chloride (0.1 g) (using a RB flask of 100 mL capacity fitted with a reflux condenser). The mixture is refluxed for 3 hr. and the solution cooled. The alkaline solution is filtered (to remove precipitated MnO_2) and SO_2 gas bubbled into the solution. Alternatively a saturated solution of sodium sulphite and dilute H_2SO_4 is added till the solution becomes colourless. The remaining solution is extracted with ether (2 × 15 mL) and ether is evaporated to give benzoic acid, which is crystallised from hot water yield 2.2 g (80-90%) M.P. 122.

Notes

1. In place of cetyltrimethyl ammonium chloride, crown ether [(18) crown 6] (0.18) can also be used. Crown ether forms a complex with $KMnO_4$, which is soluble in organic phase and then $KMnO_4$ becomes more effective for the oxidation of toluene.

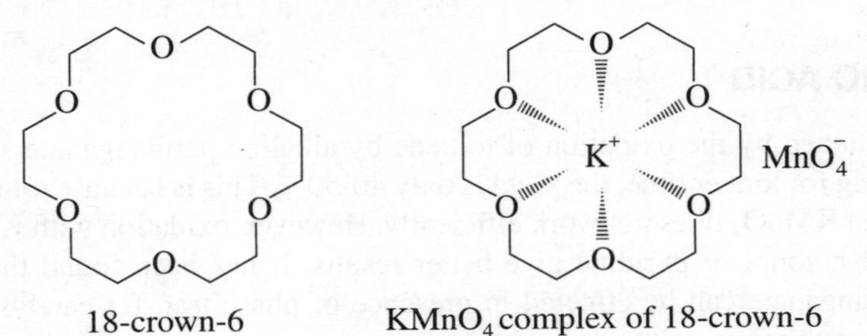

18-crown-6 — $KMnO_4$ complex of 18-crown-6

2. Benzoic acid can also be obtained by the oxidation of toluene by aq. $KMnO_4$ under microwave irradiation for 5 min in 40% yield (R. Gedye, F. Smith, K. Westaway, A. Humera, L. Baldisera, L. Laberge and Rusell, Tetrahedran Lett., 1986, **26**, 279)
3. Benzoic acid is also obtained by the oxidation of stilbene. In this case, a solution of $KMnO_4$ in water (1:10) (a suspension can also be used) is added with stirring to stilbene and tetrabutylammonium bromide. The mixture is stirred and worked up by addition of sodium bisulphite, acidification, separation and drying organic layer and evaporation. Benzoic acid is obtained in 98% yield (A.W. Harriott and D. Picker, Tetrahedron Lett., 1974, 1511).
4. Oxidation of cycloctene with alkaline $KMnO_4$ in presences of PTC gives 50% yield of cis-1, 2-cyclooctane diol compared to 7% by the classical technique (W.B. Weber and J.P. Shepherd, Tetrahedron Lett., 1972, 4907)

$$\text{Cyclooctere} + KMnO_4 \xrightarrow{C_6H_5\,CH_2\,\overset{+}{N}\,Et_3\,Cl^-} \text{Cis-1,2-Cyclooctane diol (50\%)}$$

Cyclooctere — Cis-1,2-Cyclooctane diol (50%)

5. Oxidations with PTC technique can also be carried out in solvents which are not oxidised by $KMnO_4$. Example of such solvents include t-butanel, benzene, ethylacetate, diethyl ether, acetone etc.

5.2 BENZONITRILE

A number of methods are available for the synthesis of nitriles. In most of the cases, the aldehyde is converted into oxime and then dehydrated by a wide variety of reagents.

$$RCHO \rightarrow RCH = NOH \xrightarrow{-H_2O} RCN$$

Nitrites can also be prepared by treatment of alkyl halides with KCN in presence of a PTC (see Section 5.2a). However, this procedure is not suitable for the preparation of benzonitrile.

It has been found that benzonitrile can best be prepared form benzamide on reaction with dichlorocarbene generated in situ by the reaction of chloroform and sodium hydroxide in presence of a PTC.

$$\underset{\text{Benzamide}}{C_6H_5CONH_2} + \underset{\text{Chloroform}}{CHCl_3} + \underset{\text{aq}}{NaOH} \xrightarrow{C_6H_5CH_2N^+ Et_3Cl^-} C_6H_5CN$$

Materials

Benzamide	6.05 g
Benzyl triethyl ammonium chloride	0.34 g
Chloroform	6 mL
aq. Sodium hydroxide (50%)	25 g NaOH in 50 mL H_2O

Procedure[1]

A mixture of benamide (6.05 g), benzyl triethyl ammonium chloride (0.34 g), Chloroform (6 mL) and aq. sodium hydroxide solution (50% 25 g NaOH in 50 mL water) is stirred at room temperature for 2 hr. The mixture is extracted with chloroform, organic layer washed with water and dired (Na_2SO_4). Distillation of chloroform solution gives benzonitrile, b.p. 188-189°, yield 84%.

Notes

1. T. Saraie, K. Ishiguno, K. Kawashima and K. Morita, Tetrahadron Lett. 1971, 2121.
2. For the preparation of PTC (benzyltriethyl ammonium chloride see note 2 for in preparation of phenyl isocyanide, Section 5.13).
3. Benzonitrile can also be prepared by the dehydration of benzamide by heating with P_2O_5 in about 35% yield (V.K. Ahluwalia and R.Aggarwal, Comprehensive practical organic chemistry, Preparation and quantitative analysis, Universities Press, 2004, Page 53)
4. In an efficient procedure, the aldehyde is converted into adsorbed oximate by reaction with hydroxylamine hydrochloride and potassium fluoride on alumina under microwave activation and without a solvent. The absorbed oximate is transformed into nitrile by treatment with carbon disulphide at room temperature (D) Villemin, M. Lalaoui, A.B. Alloum, Chem. Ind., (1991, 176).

$$\text{R–CHO} \xrightarrow[\text{2) } CS_2\text{, 20°, 20 hr}]{\text{1) } NH_2OH.HCl \text{ on } Al_2O_3\text{-KF, microwave 350 W, 5 min}} \text{RC} \equiv \text{N}$$

Using this procedure a number of nitrites can be prepared in good yield.

5.2A BENZYL CYANIDE

It is obtained by the reaction of benzyl chloride with sodium cyanide in presence of a PTC (benzyl trimethylammonium chloride)

$$\underset{\text{Benzyl chloride}}{C_6H_5CH_2Cl} \xrightarrow[\text{1.5 hr. 80-85°}]{\text{PTC}} \underset{\text{Benzylcyanide}}{C_6H_5CH_2CN}$$

Materials

Benzyl chloride	5 g
Sodium cyanide	3 g
Benzyl trimethyl ammonium chloride	5 g

Procedure

A mixture of benzyl chloride (5 g), sodium cyanide (3 g), benzyl trimethylammonium chloride (5 g) and water (10 mL) is heatest at 80-85° for 1.5 hr. The reaction mixture is cooled and extracted with ether. Ether is removed by distillation to give benzyl cyanide, b.p. 161-62°/100 mm. yield 4 g (90%)

Notes

1. Using 2-phenylethyl bromide in place at benzyl chloride given 3-phenylpropionitrite, b.p. 114-18°/8 mm in 90% of yield.

5.3 N-BUTYL BENZYL ETHER

Ethers are normally prepared by dehydration of alcohols

$$\text{R — OH} \cdot \xrightarrow[\text{140°}]{H_2SO_4} \text{R — O — R}$$

$$\text{(R — OH} \xrightarrow[\text{240-260°}]{Al_2O_3} \text{R — O — R)}$$

This procedure is useful for the preparation of simple ethers only. However mixed ethers like n-butyl benzyl ether can be obtained by the reaction of alkoxides or phenoxides with alkyl halides. This method is known as **Williamson's Ether synthesis**[1]. However the yield are low.

$$RO^- + R'—X \longrightarrow R—O—R'$$

$$\underset{\text{Sod-salt of benzyl alcohol}}{C_6H_5CH_2ONa} + \underset{\text{n-Butyl chloride}}{ClCH_2CH_2CH_2CH_3} \xrightarrow{\Delta} \underset{\text{n-Butyl benzyl ether}}{C_6H_5CH_2OCH_2CH_2CH_2CH_3}$$

It has been found that Williamson ether synthesis proceeds very well in presence of a PTC. Thus, the reaction of butyl alcohol with benzyl chloride in presence of sodium hydroxide and tetrabutylammonium bisulphate (TBAB) as catalyst gives n-butyl benzyl etter in excellent yields.

$$\underset{\text{n-Butyl alcohol}}{CH_3CH_2CH_2CH_2OH} + \underset{\text{Benzyl chloride}}{C_6H_5CH_2Cl} \xrightarrow[\text{TBAB}]{\text{50\% Na OH}} \underset{\text{n-Butyl benzyl ether}}{CH_3CH_2CH_2CH_2OCH_2C_6H_5} + NaCl + H_2O$$

Materials

n-Butyl alcohol	6.2 g
Benzyl chloride	6.3 g
Tetrabutylammonium bisulphate (TBAB)	1 g
Sodium hydroxide 50%	10 g in 20 mL H_2O

Procedure[2]

A mixture of n-butyl alcohol (6.2 g), benzyl chloride (6.3 g), sodium hydroxide solution (50%, 10 g in 20 mL H_2O) and TBAB (1 g) is stirred at 35-40° for 1.5 hr. The reaction mixture is extracted with THF, the THF extract washed with sat'd solution of sodium chloride, dried (Na_2SO_4) and distilled to give n-butyl benzyl ether in 92% yield.

Notes

1. A.W. Williamson, J. Chem. Soc., 1982, **4**, 229; O.C. Dermer, Chem. Revs., 1934, **14**, 409.
2. H.H. Freedman and R.A. Dubois, Tetrahedron Lett. 1975, 3251.
3. This is the best procedure for the preparation of mixed ethers. The usual Williamson method gives a mixture of products.
4. In this procedure, only a minor amount of symmetrical ether is formed.

   Primary alcohol are completely alkylated by aliphatic chlorides; secondary alcohols require longer time or greater amount of catalyst.
5. Mixed ethers can also be obtained by treatment of alcohol with halides in presence of sonication using PEG as solvent. The yields are 80% compared to 44% without sonication. Thus ethyl phenyl ether is prepared as follows (R.S. Davidson, A. Safdar, J.D. Spencer and D.W. Lewis, Ultrasonic, 1987, **25**, 35

$$C_2H_5OH + C_6H_5X \xrightarrow[)))]{KOH,\ PEG} \underset{80\%}{C_2H_5OC_6H_5}$$

6. We know that dimethyl sulphate does not react with most alcohols in the presence of aqueous sodium hydroxide or even by the use of alkali metal alkoxides the reaction proceeds easily with tetrabutylammonium salts as catalyst (A. Merz, Angew. Chem. Int. Ed. Engl., 1973, **12**, 846).
7. Ethers from phenols are obtained by using crown ethers as PTC. Thus, benzyl phenyl ether is synthesised in quantitative yield as follows.

$$\underset{\text{Phenol}}{C_6H_5OH} + \underset{\text{Benzyl chloride}}{C_6H_5CH_2Cl} \xrightarrow[\text{Crown ether [18-Crown-6]}]{K_2CO_3} \underset{\text{Benzyl phenyl ether}}{C_6H_5OCH_2C_6H_5}$$

5.4 1–CYANOOCTANE

It is known that alkyl halides (e.g., 1-chlorooctane) do not react with sodium cyanide under a variety of conditions, viz., stirring and heating for long time (in some cash however very poor yield of the corresponding nitrile in obtained). It is, however, found that, if a small quantity of a PTC in used, the reaction goes to completion in shorter time (about 2 hr) to give the corresponding nitrile.

$$\underset{\text{1-Chlorooclane}}{CH_3(CH_2)_6CH_2Cl} + \underset{\text{or KCN}}{NaCN} \xrightarrow[C_{16}H_{33}\overset{+}{P}(C_4H_9)_3\overset{-}{Br}]{PTC} \underset{\substack{\text{1-Cyanooctane} \\ 94\%}}{CH_3(CH_2)_6CH_2CN}$$

Materials

1-Chlorooctane	5 g
Sodium cyanide	5 g
Water	5 mL
Hexadecyltributyl phosphonium bromide (PTC)	0.5 g

Procedure

A mixture of 1-chlorooctane (5 g), Sodium cyanide (5 g), water (5 mL) and hexadecyltributyl phosphonium bromide (0.5 g) is heated with stirring at 105° for 2 hrs in a R.B flask (250 mL capacity). The reaction mixture is cooled, water (100 mL) added and extracted with dichloroe thane, The organic extract it is washed with water, dried (Na_2SO_4) and distilled to give 1-cyanooctane, b.p. 140°/1mm yield 94%.

Notes

1. C.M. Stark, J.Am. Chem. Soc., 1971, **93**, 195
2. The PTC, hexadecyl tributyl phosphonium bromide [n–$C_{16}H_{33}P^+(n–C_4H_9)_3\,Br^-$] is prepared (CM. Starks. J. Am. Chem. Soc., 1971 **93**, 195) by heating a mixture of 1-bromohexadecane (1.5 g) and tributyl phosphine (1 g) at 70-90° for 3-4 days. Cool the mixture, filter the separated PTC and crystallise from hexane. Yield 1.7 g (60%), m.p. 54°.
3. PTC methodology can also be used for the preparation of following alkyl or acyl halides

(i) $$\underset{\text{2-Phenylethyl bromide}}{C_6H_5CH_2CH_2Br} + \underset{\text{aqueous}}{NaCN} \xrightarrow[\text{3 hr., 95\%}]{\text{PTC},\; N^+(CH_3)_3CH_2C_6H_5\bar{Cl}} \underset{\text{3-Phenylpropionitrile, 91\%, bp. 114-18°/8mm}}{C_6H_5CH_2CH_2CN}$$

Ref. N. Sugimoto, T. Fujita, N. Shigematsu and A. Ayada, Chem. Pharm. Bull. 19 62, **10**, 427; Japanese Patent 1961/63.

(ii) $$\underset{\text{Benzoyl chloride}}{C_6H_5\overset{O}{\overset{\|}{C}}Cl} + \underset{\text{aq}}{NaCN} \xrightarrow[Bu_4\overset{+\,-}{N}X]{\text{PTC}} \underset{\text{Benzoyl cyanide, 60-70\%}}{C_6C_5\overset{O}{\overset{\|}{C}}—CN} + NaCl$$

Ref. K.E. Koening and W.P. Weber, Tetrahedron Lett., 1974, 2275; H. Habner, Ann. Chem. Pharm, 1861, **120**, 33.

4. Benzonitrite cannot be prepared by the above procedure. However, it can be conventiently prepared from Benzamide by the reaction with dichlorocarbene (generated in situ) (See Section 5.2).

5.5 DICHLORONORCARANE [2,2-DICHLOROBICYCYO (4.1.0) HEPTANE]

It is obtained by the addition of dichlorocarbene genaterated in alkaline medium from chloroform). This reaction was earlier conducted in aq. NaOH solution; in this case the required product was obtained only in 0.5% yield. This is because, the generated dichlorocarbone undergo hydrolysis in aqueous medium.

$$\text{Cyclohexene} + CHCl_3 + \underset{\text{aq.}}{NaOH} \longrightarrow \underset{0.5\%}{\text{7,7-dichloronorcarane (Cl, Cl)}}$$

$$CHCl_3 + \text{base} \longrightarrow C\bar{Cl}_3 \xrightarrow{-\bar{Cl}} :CCl_2 \xrightarrow{H_2O} CO + HCO\bar{O} + \bar{Cl}$$

However, if sodium ethoxide or potassium tert. butoxide is used as a base and the reaction is conducted in anhydrous solvent, the required dichloronorcarane is obtained in 60-70% yield.

$$\text{Cyclohexene} + CHCl_3 \xrightarrow[\text{anhyd}\cdot\text{solvent}]{\text{NaOEt or pot. t-butoxide}} \text{7,7-dichloronorcarane (60-70\%)}$$

It has been found that by the use of a PTC, the addition product is obtained in 60-70% yield.

$$\text{Cyclohexene} + CHCl_3 + \text{aq. NaOH} \xrightarrow[\text{Benzyltriethyl ammonium chloride}]{\text{PTC}} \text{7,7-dichloronorcarane (60-70\%)}$$

Better yields in the case of PTC is because $:CCl_2$ generated in situ is transferred to the organic phase as soon as it is generated and reacts with cyclohexene than reacting with water.

Materials

Chlroform	12 mL
Cyclohexene	4.1 g (5.1 mL)
Tetra-n-butyl ammonium bromide	0.25 g
Sodium hydroxide	20 mL (50%)

Procedure

To a stirred mixture of cyclohexene (5.1 mL) and tetra-n-butyl ammonium bromide (0.25 g) contained in RB flask (50 mL capacity, fitted with a reflux condenser) is slowly added chloroform (12 mL). The stirred mixture is treated with sodium hydroxide solution (50% 20 mL) and water (15 mL) added in one lot. The stirring is continued for 30 min while the reaction mixture is gently heated. The reaction mixture is cooled to room temperature and extracted with ether (2 × 30 mL). The ether extract is dired (Na_2SO_4) and distilled to dicholonorcarane, b.p. 192-97° yield 5.7 g (65%)

Note

1. W. Vone., Doering and A.K. Hoffmann J, Am. Chem. Soc., 1954, **76**, 612.
2. W.E. Parham and E.E Schweizer, Org. Rect. 1963, **13**, 55
3. A.P. Kreshkov, E.N. Sugushkima and B.A. Krozdov, J. Appl. Chem. USSR (Engl. Transl.), 1965, **38**, 2357

4. Using styrene in place of cyclohexene it in possible to prepare 1-phenyl-2, 2-dichlorocyclopropane (b.p. 103°/10 Torr) (M. Makosza and M. Wawrziewicz, Tetrahedron Lett, 1969, 4659).

C_6H_5, H, H, H, Cl Cl

5. Dichlorocarbene can also be generated by the direct reaction (S.L. Regen and A. Singh, J. Org. Chum. 1982, **47**, 1587) between powdered sodium hydroxide and chloroform by soniction. This procedure is simple and avoids the use of PTC. The in situ generation of dichlorocarbene undergoes addition to alkenes. Thus, sonication of styrene with solid sodium hydroxide and chloroform by stirring gives 96% yield of the adduct in 1 hr.

Styrene $\xrightarrow[\text{)))), stirring}]{\text{NaOH, CHCl}_3}$ (Cl Cl adduct)

6. **Reimer-Tiemann reaction** can also be performed using the same procedure see preparation of salicylaldehyde (Section 5.14)

5.6 4, 6-DIMETHYL-3-PHENYLCOUMARIN

3-Arylcumarnis are used as optical brightners (A. Dovlars, C.W. Schellhammer are J. Schroeder, Angew Chem. Internal. Edn., 1975, **14**, 665). These were earlier obtained in low yields and required anhydrous conditions (P. Pulla Rao and G. Srimannarayana, Synthesis, 1981, **11**, 887, T.R. Seshadri and S.Varadarajan J. Sci. Ind. Res., 1952, **Section B, 11**, 48; Proc. Indian Acad. Sci., 1952, **Sect. A, 35**, 75), It has been found that these cumarins can be obtained in excellent yield and high purity by using a PTC catalyst in presence of aqueous potassium carbonate by the reaction of o-hydroxycarbonyl compounds with phenyl acetyl chloride (G. Sabitha and A.V. Subha Rao, Synth. Communication, 1987, **17(3)**, 341)

(OH, H_3C, $COCH_3$) + C_6H_5 CH_2COCl $\xrightarrow[\text{aq. } K_2CO_3]{\text{Tetrabutyl ammonium hydrogen sulphate}}$ (O, O, H_3C, C_6H_5, CH_3)

4,6-Dimethyl-3-phenylcoumarin

Materials

2-Hydroxy-5-methylacetophenene	2.5 g
Phenylacetyl chloride	1.5 mL
Tetrabutylammonium hydrogen sulphate	100 mg

Procedure

A solution of phenylacetyl chloride (1.5 mL) in benzene (10 mL) is added dropping during 10-15 mm to a stirred mixture of 2-hydroxy-5-methyacetophenone (2.5 g) in benzene (25 mL), tetrabutylammonium hydrogen sulphate (100 mg) and aqueous potassium carbonate (20%, 30 mL). The mixture is stirred for about 3 hr. at room temperature. The organic layer is separated, washed with water, dried (Na_2SO_4) and distilled The residual product is crystallized from ethanol, m.p. 172° yield 96%.

Note

1. A number of other 3-alkylcoumarins having various oxygenation pattern and substituted in the phenyl group can be synthesed by the above PTC procedure in good yield.

5.7 2,2 DIMETHYL-3-PHENYL PROPION ALDEHYDE

Aldehydes containing only an α-hydrogen atom, such as isobutyraldehyde can be alkylaled with alkyl halides in the presence of 50% aqueous sodium hydroxide and a catalytic amount of tetrabutylammonium ions (H.Dietl and K.C. Brannock, Tetrahedran Lett., 1973, 1273).

$$\underset{\text{Isobutyraldehyde}}{(CH_3)_2CHCHO} + \underset{\text{Benzyl chloride}}{C_6H_5\,CH_2Cl} + \underset{\text{aq}}{NaOH} \xrightarrow{Bu_4\overset{+}{N}\overset{-}{Y}} C_6H_5\,CH_2-\underset{CH_3}{\overset{CH_3}{\underset{|}{\overset{|}{C}}}}-CHO$$

2,2-Dimethyl-3-phenyl propionaldehyde 75%

Materials

Isobutyraldehyde	9.8 g
Benzyl chloride	12.66 g
Sodium hydroxide solution	4.66 g in water 5 mL
Tetrabutylammonium iodide	0.49 g

Procedure

Aqueous sodium hydroxide solution (4.66 g in 5 mL water) is added dropwise during 2-3 hr to stirred mixture of isobutyraldehyde (9.8 g, 0.133 mol), benzyl chloride (12.66g, 0.1 mol), benzene (10 mL) and tetrabatylammonium iodide (0.49 g) at 70°. The reaction mixture is stirred at 78 for

2 hr. more. The organic phase is separated, washed with water, dried (Na_2SO_4) and distilled to give 2, 2-dimethyl-3-phenyl propionaldehyde, b.p. 95°/17.2 mm. yield 12.1 of (75%)

Note

1. This is a general procedure for the alkylation's of aldehydes containing only an α-hydrogen atom.

5.8 3,4-DIPHENYL-7-HYDROXYCOUMARIN

3, 4-Diphenyl coumarins, known for their antifertility activity, were earlier prepared by the Perkin reaction[1] involving heating o-hydroxybenzophenones with sodium phenylacetate in acetic anhydride; the yields in this reaction are very poor. These are now obtained in excellent yield by using a PTC in presence of aqueous potassium carbonate by the reaction of o-hydroxybenzophenones with phenyl acetyl chloride.

CH_3O, OH, CO, C_6H_5 + $C_6H_5CH_2COCl$ (Phenyl acetyl chloride) $\xrightarrow[\text{aq. } K_2CO_3\text{, benzene}]{\text{tetrabutyl ammonium hydrogen sulphate}}$ HO, O, O, C_6H_5, C_6H_5

2-Hydroxy-4-methoxy benzophenone

3,4-Diphenyl -7-hydroxycoumarin

Materials

2-Hydroxy-4-methoxy benzophenone	2.45 g
Phenyl acetyl chloride	1.5 mL
Tetrabutyl ammonium hydrogen sulphate	0.1 g
Aqueous potassium carbonate 20%	50 mL

Procedure[2]

Phenyl acetylchloride (1.5 mL) in benzene (10 mL) is added dropwise to a stirred mixture of 2-hydroxy-4-methoxybenzophenone (2.45 g) in benzene (50 mL), tetrabutyl ammonium hydrogen sulphate (0.1 g) and aqueous potassium carbonate (50 mL, 20%). The mixture is stirred for 5 hr, organic layer separated, washed with water, dried (Na_2SO_4) and solvent removed (distillation). The residual product is crystallised from ethanol to give 3, 4-diphenyl – 7-hydroxycoumarin, m.p. 168-69° yield 80%.

Note

1. W.H. Perkin, J. Chem. Soc., 1968, 21, 58, 181; 1877, **31**, 388; J.R. Johnson, Org. Reactions, 1942, 1 210.
2. V.K. Ahluwalia and C.H. Khanduri, India J. Clem., 1989, **28 B**, 599.
3. The PTC, tetrabutyl ammonium hydrogen sulphate is obtained (W.T. Ford and R.J. Haurt, J.Am. Chem. Soc. 1973 **95**, 7381) as follows.

   Dimethyl sulphate (4.6 g) is added to a stirred mixture of tetrabutyl ammonium bromide (9.8 g) and chlorobenzene (15 mL) at 80-85° in a two neck round bottomed flask fitted with a short distillation column and a dropping funnel. The formed methyl bromide is collected as a distillate using a trap cooled in acetone-dry ice mixture. After the distillation of methyl bromide ceases, the heating is increased until the temperature at the top of the distillation column starts to rise rapidly. A solution of concentrated sulphuric acid (0.75 mL) in water (300 mL) is added cautiously and the mixture refuxed for 48 hr. The solution is evaporated to almost dryness under reduced pressure. The residue in dichloromethane (250 mL), is washed with water (2 × 30 mL), dried (anhyd-sodium sulphate) and distilled the solvent. The PTC, Tetrabutylammonium hydrogen sulphate (10g) separates out. It is almost pure and can be crystallised from isobutyl methyl ketones.
4. The above PTC method is very convenient for the synthesis at 3, 4-disubstituted coumarins (G. Sabitha and A.V. Subba Rao, Synth. Common., 1987, **17**, 341). In this method an o-hydroxycarbonyl compound is reacted with an appropriately substituted phenylacetyl chloride in presence of aqueous K_2CO_3 using a PTC as a catalyst

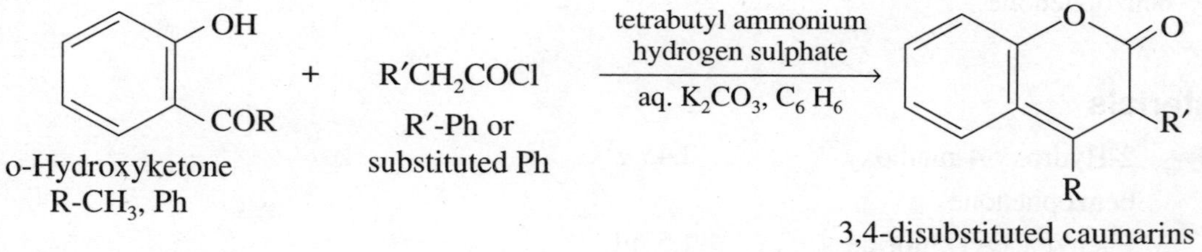

OH, COR (o-Hydroxyketone, R-CH_3, Ph) + $R'CH_2COCl$ (R′-Ph or substituted Ph) $\xrightarrow[\text{aq. } K_2CO_3,\ C_6H_6]{\text{tetrabutyl ammonium hydrogen sulphate}}$ O, O, R′, R

3,4-disubstituted caumarins

Use of phenacetyl bromide in place of phenaylacetyl chloride in the above procedum gives [2-benzoyl-3, 5-dimethyl benzofuran, m.p 15° in 70% yield]

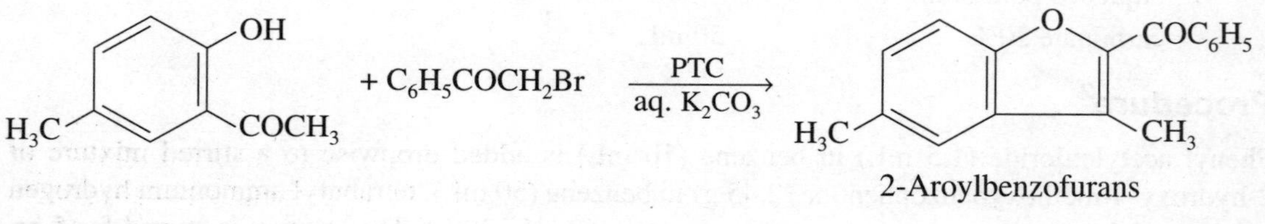

OH, H_3C, $COCH_3$ + $C_6H_5COCH_2Br$ $\xrightarrow[\text{aq. } K_2CO_3]{\text{PTC}}$ O, COC_6H_5, H_3C, CH_3

2-Aroylbenzofurans

5.9 FLAVONE

Flavones, a class of natural products where synthesed earlier by a number of methods in low yields and the work up procedure is difficult. A convenient and general method of synthesis is the **Baker-Venkataraman rearrangement**. In this procedures, an ortho-benzoylacetophenone is isomerised to an ortho-hydroxy-β-diketone with a base. This rearrangement occurs by an **internal claisen condensation**. Treatment of the formed β-diketone with acid gives the flavone.

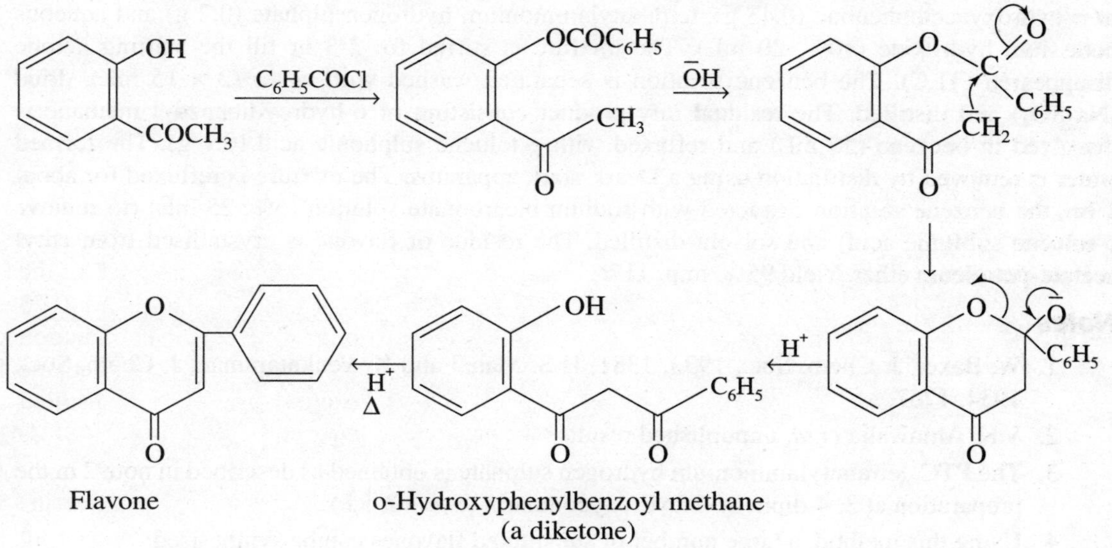

Flavone o-Hydroxyphenylbenzoyl methane (a diketone)

It has been found that in presence of a PTC, in alkaline solution gives much better yield of the o-hydroxydibenzoyl methane

o-hydroxydibenzoyl methane

flavone

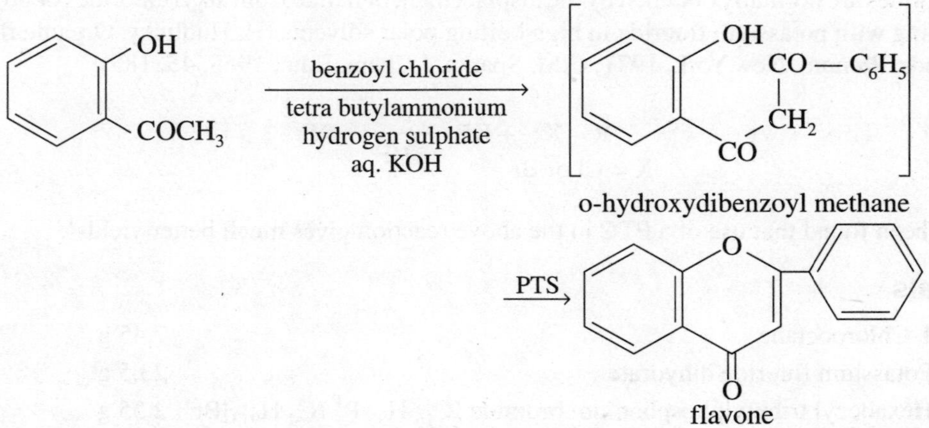

Materials

o-Hydroxyacetophenone	0.43 g
Benzoyl chloride	0.42 g
Tetrabutyl ammonium hydrogen sulphate	0.2 g
Aq. Potassium hydroxide 10%	20 mL
p-Toluene sulphonic acid	0.1 g

Procedure[2]

A solution of benzoyl chloride (0.42 g) in benzene (20 mL) is added slowly to a stirred mixture of o-hydroxyacetophenone (0.43 g), tetrabutylammonium hydrogensulphate (0.2 g) and aqueous potassium hydroxide (10%, 20 mL). The mixture is stirred for 2-3 hr till the starting ketone disappeared (TLC). The benzene solution is separated washed with water (2 × 15 mL), dried (Na_2SO_4) and distilled. The residual oily product consisting of o-hydroydibenzoyl methane is dissolved in benzene (50 mL) and refluxed with p-toluene sulphonic acid (0.1 g). The formed water is removed by distillation using a Deark-stark apparatus. The mixture is refluxed for about 1 hr., the benzene solution extracted with sodium bicarbonate solution (5%, 25 mL) (to remove p-toluene subfonic acid) and solvent distilled. The residue of flavone is crystallised from ethyl acetate-petroleum ether. Yield 95%, m.p. 117°.

Notes

1. W. Baker, J. Chem. Soc., 1933, 1381; H.S. Mahal and K. Venkataraman, J. Chem. Soc., 1934, 1767.
2. V.K. Ahluwalia *et al.* unpublished results.
3. The PTC, tetrabutylammonium hydrogen sulphate is obtained as described in note 2 in the preparation at 3, 4-dipheny-7-hydroxycoumarin (Section 5.8)
4. Using this method, a large number of substituted flavones can be synthesised.

5.10 1-FLUOROOCTANE

Alkylfluorides are normally obtained by the displacement of halide from alkyl chlorides or bromides by refluxing with potassium fluoride in high boiling polar solvents (H. Hudlicky, Organic fluorine compounds, Plenum, New York, 1971; C.M. Sparks, J.Chem. Edu., 1968, **45**, 185).

$$R-X \quad + KF \xrightarrow[THF]{\Delta} RF$$
$$X = Cl \text{ or } Br$$

It has been found that use of a PTC in the above reaction gives much better yields[1].

Materials

1-Chlorooctane	7.45 g
Potassium fluoride dihydrate	23.5 g
Hexadecyl tributylphosphonium bromide $[C_{16}H_{33}P^T(C_4H_9)_3]Br^-$	2.55 g

Procedure

A stirred mixture of 1-chloroctane (7.450, 0.05 mol), potassium fluoride dihydrate (23.5 g, 0.25 mol), hexadecyl tributylphosphonium bromide (2.55 g, 0.005 mol) and water (15 mL) is heated at 150-160° for 6-7 hr. The organic layer is separated, washed with water, conc. H_2SO_4 and again with water. It is dried ($CaCl_2$) and distilled to give 5.1 g (77%) 1-fluoroctane, b.p. 142-144°.

5.11 1-OXASPIRO-[2, 5]-OCTANE-2-CARBONITRILE

It is normally obtained by **Darzen condensation**[1] involving the condensation of aldehydes and ketones (cyclohexanone) with chloroacetonitrile in presence of a base, which is generally lithium bis (trimethyl silyl amide, [Li N (Si $Me_3)_2$] and the reaction is performed in THF at-78°.

$$ClCH_2CN \xrightarrow{base} Cl\bar{C}HCN$$

Chloroaceto nitrile

$$\text{Cyclohexanone} + Cl\bar{C}HCN \longrightarrow \text{(alkoxide intermediate)} \xrightarrow{-Cl^-} \text{1-Oxaspiro-[2-5]-octane-2-carbonitrile}$$

Cyclohexane 1-Oxaspiro-[2-5]-octane-2-carbonitrile

It is found that the above reaction proceeds very well in presence of a PTC[2] (benzyl triethyl ammonium chloride) in aqueous sodium hydroxide

Materials

Cyclohexanone	5.4 g
Sodium hydroxide solution	10 mL (5%)
Benzyltriethyl ammonium chloride (PTC)	0.2 g
Chloroacetonitrile	3.8 g

Procedure

Chloroacetonitrile (3.8 g) is added dropwise is a stirred mixture of cydohexanone (5.4 g), sodium hydroxide solution (5%, 10 mL) and benzyltriethylammonium chloride (0.2 g). The mixture is stireed for 30 min. and extracted with dichloro methane. The organic extract is washed with water, dried (Na_2SO_4) and distilled to give 1-oxaspiro-[2,5]octane-2-carbonitrile, b.p.87°/5 mm. yield 70%.

Notes

1. C. Darzens, Comp. Rend., 1904, **139**, 1214; 1905, **141**, 766; 1906, **142**, 214; M. Ballester, Chem. Rev., 1955, **55**, 283.
2. A. Joncyk, M. Fedorynski and M. Makosa, Tetraedran Lett., 1972, 2395:
3. Normally Darzen condensation involves the condensation of aldehydes or ketones with α-halester in presence of a base to yield α, β-epoxyesters called glycidic esters[1].

$$R-\overset{O}{\overset{\|}{C}}-R' + ClCH_2COOC_2H_5 \xrightarrow{base} \text{RR'C}\underset{O}{-}\text{CH}-COOC_2H_5$$

Glycidic ester

4. The required PTC benzyltriethylammonium chloride is prepared as described in the preparation of phenylisoyanide (see Section 5.13, note 2)
5. In Darzen condensation with aldehydes and unsaturated ketones, both possible stereoisomers are obtained. However, with more acidic ketones like phenyl acetone, the ketone carbanion is formed rather than the nitrile, leading to the alkylation of the ketone.

$$C_6H_5CH_2COCH_3 + ClCH_2CN + \underset{aq}{NaOH} \xrightarrow[Q^+X^-]{PTC} C_6H_5\underset{\displaystyle CH_2CN}{\underset{|}{C}}HCOCH_3$$

5.12 2-PHENYLBUTYRONITRILE

Alkylation of nitriles with alkyl halides takes place in the α-prosition and requires strong bases like sodium amide, metal hydride, potassium tertiary butoxide and involve use of anhydrous organic solvents (A.C. Cope, H.L. Holmes and H.O. House, org. React., 1957, **9**, 107) and the yields are low.

$$C_6H_5CH_2CN + C_2H_5Cl \xrightarrow[Toluene]{NaNH_2} C_6H_5\underset{\displaystyle C_2H_5}{\underset{|}{C}}HCN$$

It has been found that due to high selectivity of PTC, it is used for the synthesis of manoalkyl derivatives of nitriles (M. Makosza, Tetrahedron, 1968, **24**, 175). The reaction is conducted in aqueous alkali.

$$C_6H_5CH_2CN + C_2H_5Cl \xrightarrow{C_6H_5CH_2\overset{+}{N}(C_2H_5)_3\overset{-}{Cl}} C_6H_5\underset{\displaystyle C_2H_5}{\underset{|}{C}}HCN + NaCl$$

80-85%

Materials

Phenyl acetonitrite	12.85 g
Sodium hydroxide solution	50%, 27 mL
Benzyl triethylammonium chloride	0.25 g
Ethyl bromide	10.9 g, 7.5 mL
Benzaldehyde	1.06 g, 1.01 mL

Procedure

Ethyl bromide (7.5 mL, 0.1 mL) is added dropwise during 60 min. to a stirred mixture of aqueous sodium hydroxide (50% 27 mL), phenyl acetonitrite (12.85 g, 0.11 mol) and benzyl triethylammonium chloride (0.25 g, 0.011 mol) taken in a three necked RB flask fitted with a mechanical stirrer and reflux condenser; the temperature during the addition of ethyl bromide is kept at 28-35° (if necessary, cold water is used to cool the reaction flask). After the addition is over, the mixture is stirred at 40° for 30 min more. The mixture is cooled (25°) and distilled benzaldehyde (1.06 g, 1.01 mL, 0.01 mol) is added and stirring continued for 60 min. Benzaldehyde reacts with unrealed phenylacetonitrite to form high boiling α-phenylcinnamonitrile. The reaction mixture is cooled (ice-cold water), water (38 mL) and benzene (10 mL) added. The organic layer is separated and the aqueous phase is extracted with benzene (2 × 10 mL). The combined benzene extract is washed with water (2 × 10 mL), dil. HCl (5%, 2 × 10 mL) and water (2 × 10 mL). The organic layer is dried (Mg SO_4) and distilled to give 2-phenylbutyronitrite, b.p. 102-4°/7 mm. yield 12.1 g (84%).

Notes

1. Using the same procedure a number of α-substituted nitrites can be obtained. Some examples are gives below.

$$C_6H_5CH_2CN + (C_6H_5)_2CHCl \xrightarrow[\text{aq.NaOH}]{\text{PTC}} (C_6H_5)_2CHCH(C_6H_5)CN \quad 94\%$$

$$C_6H_5CH(C_2H_5)CN + C_6H_5CH_2Cl \xrightarrow[\text{aq.NaOH}]{\text{PTC}} C_6H_5CH_2C(C_6H_5)(C_2H_5)CN \quad 94\%$$

5.13 PHENYLISOCYANIDE ($C_6H_5\,N \equiv C$)

Generally isocyanides are prepared by the **Hoffmann carbylamine reaction**[1] involving the reaction of primary amines with chloroform in presence of alkali.

$$RNH_2 + CHCl_3 + 3NaOH \rightarrow RN \equiv C + 3NaCl + 3H_2O$$

The mechanism of the reaction is given below

$$CHCl_3 + \bar{O}H \longrightarrow \bar{C}Cl_3 + H_2O$$

$$\bar{C}Cl_3 \longrightarrow :\bar{C}Cl_2 + \bar{Cl}$$

Dichlorocarbene

$$R\dot{N}H_2 + :CCl_2 \longrightarrow RNH\,CHCl_2$$

$$RNH\,CHCl_2 + 2\bar{O}H \longrightarrow R-\overset{+}{N}\equiv\bar{C} + 2\bar{Cl} + 2H_2O$$

However, the yields in this reaction are poor (5-10%). It is found that the use of a PTC in the above reaction gives very good yield.

$$CHCl_3 + NaOH \xrightarrow{PTC} :CCl_2 \xrightarrow{C_6H_5NH_2} \underset{57\%}{C_6H_5N{\equiv}C}$$

Materials

Aniline	4 mL
Chloroform	12 mL
aq. NaOH (50% solution)	20 mL
Benzyl Triethylammonium Chloride (PTC)[2]	0.25 g

Procedure

To a vigorously stirred sodium of aqueous sodium hydroxide (20 mL, 5%) is added a mixture of aniline (4 mL) and benzyltriethlammonium chloride[2] (0.25 g) (using a flask fitted with reflux condenser). When the reaction subsides (about 5 min), the stirring is continued for 1 hr. The mixture is extracted with methylene chloride, organic layer washed with aqueous sodium chloride (5%, 30 mL), dried (Na_2SO_4) and distilled (under nitrogen atmosphere) to give phenyl isocyanide, b.p. 50-52°/11 mm. Yield 57%.

Notes

1. The reaction is also known as **carbylamin reaction** (A.W. Hofmann, Ann. 1868, **146**, 107; Bew, 1870, **3**, 767)
2. The required PTC, viz. benzyl triethylammonium chloride is obtained by refluxing a solution of triethlamine (3.3 g) and benzyl cholride (5 g) in absolute ethanol for about 50 hrs. The solution is cooled to room temperature and ether added. The separated benzyltriethylammonium chloride is filtered and purified by dissolving in hot acetone and reprecipitation with ether. Yield 8.5 g (90%).
3. The reaction using PTC is known as **Phase transfer Hofmann carbylamine reaction.**
4. Using the above PTC procedure following isocyanides have been prepared (W.B. Weber and G.W. Gukil, Tetrahedrm lett., 1972, 1637)

Isocyanides	*% yield*	*B.P.*
$(CH_3)_3\,CN \equiv C$	73	92-3°/722 mm
$CH_3(CH_2)_2\,CH_2\,N \equiv C$	60	40-2°/11 mm
$C_6H_5CH_2\,N \equiv C$	45	92-3°/11 mm
$CH_3(CH_2)_{10}\,N \equiv C$	41	115-18°/0.1 mm
$C_6H_5\,N \equiv C$	57	50-2°/11 mm
$CH_3\,N \equiv C$	24	59-60°/760 mm
$CH_3\,CH_2\,N \equiv C$	47	78-9°/760 mm

For the synthesis of last two isocyanides bromoform is used in place of chloroform due to ease of fractionation (W.B. Weber, G.W. Gokil and I.Ugi Angew. Chen. Int. Ed. Engl., 1972, **11**, 530).

5. In the above Phase Transfer Hofmann carbylamine reaction, use of secondary amine in place of primary amines, gives the corresponding N, N-disubstituted formamides (J. Graefe, I. Forehlick and M. Muehlstoedt, Z. Chem., 1974, **14**, 434; M. Makosza and A. Kacprowicz, Rocz. Chem., 1975, **49**, 1627)

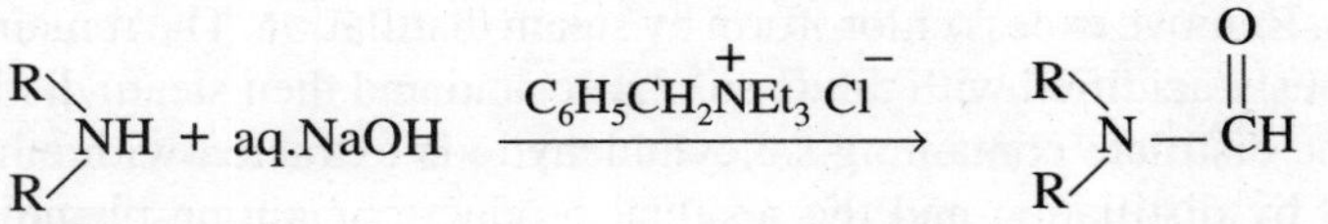

$$R_2NH + \text{aq.NaOH} \xrightarrow{C_6H_5CH_2\overset{+}{N}Et_3\ \overset{-}{Cl}} R_2N-\overset{O}{\overset{\|}{C}}H$$

Sec amine — (85%)

R = R = ethyl, 2-butyl, cyclohexyl, allyl — N,N-Disubstituted farmamides

5.14 SALICYALDEHYDE

It is obtained by the reaction of phenol with chloroform in the presence of sodium hydroxide. This reaction is known as **Reimer-Tiemann Reaction**. However, the yield of salilyaldehyde is low (about 25%) Various steps involved in the reaction are given below.

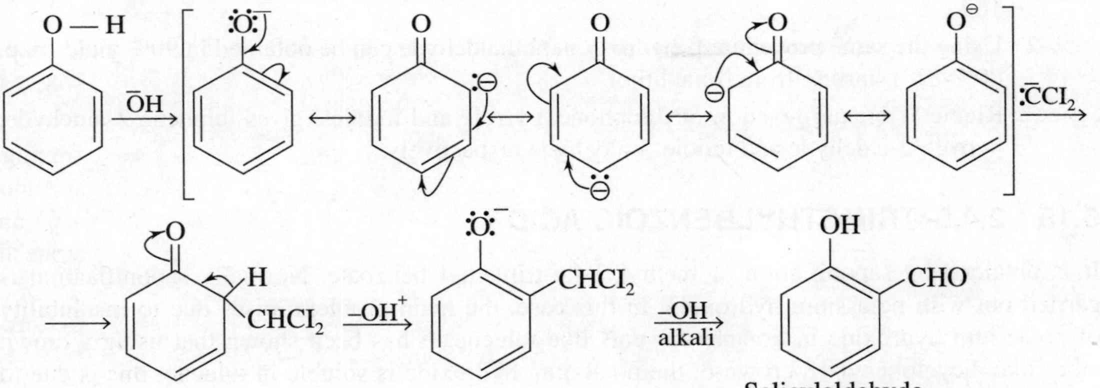

$$C_6H_5OH \xrightarrow{\bar{O}H} [\text{phenoxide ion} \longleftrightarrow \text{resonance structures}] \xrightarrow{:\bar{C}Cl_2} \text{cyclohexadienone with H and } CHCl_2 \xrightarrow{-OH^+} o\text{-}(CHCl_2)C_6H_4O^- \xrightarrow[\text{alkali}]{-OH} o\text{-}HOC_6H_4CHO$$

Salicylaldehyde

In Reimer Tiemann formylation, the formyl group is directed to ortho-position unless one of the ortho-positions or both are occupied, in that case the attack is para. Thus, in the above reaction, along with saticyaldehyde, a minor amount of p-hydroxy benzaldehyde is also obtained.

The yield of salicylaldehyde is considerably increased if the above reaction is carried out in presence of a PTC catalyst

Materials

Phenol	16.3 g
Chloroform	27.3 mL
Sodium hydroxide	50% 20 mL
Tetra-n-butyl ammonium bromide	0.25 g

Procedure

To a stirred mixture of chloroform (27.3 mL), phenol (16.3 g) and tetranbatylammonium bromide (0.25 g) (contained in a R.B Flask, 100 mL capacity, and fitted with a reflux condenser) is added sodium hydroxide solution (50%, 20 mL) and water (15 mL). The mixture is kept stirred and heated gently for 30 min. Remove excess chloroform by steam distillation. The remaining aqueous alkaline solution is cautiously acidified with dilute sulphuric acid and then steam distilled till no more oily drops collect. The distillate containing salicylaldehyde is extracted with ether (2 × 15 mL). The ether is removed by distillation and the residual product containing phenol and salicylaldehyde solution is shaken with saturated sodium bisulphate solution and allowed to stand for 1 hr. The bisulphite adduct is filtered, washed with water and then decomposed by warming the adduct with dilute sulphuric acid. The mixture is cooled, extracted with ether, ether solution dried (Na_2SO_4) and distilled. Salicylaldehyde is collected by distillation, b.p. 195-197°. Yield 7.1 g (40%).

The p-hydroxybenzaldehyde obtained as a byproduct is isolated from the residue left after steam distillation. The solution is filtered, cooled and extracted with ether. The solvent is distilled and the residue is crystallised form aqueous sulphurous acid yield 1.5 g (7%). m.p. 116-117°.

Notes

1. K. Reimer and F. Tiemann, Ber., 1876, **9**, 824, 1205, H. Wynberg, Chem. Rev. 1960, **60**, 169.
2. Using the same procedure β-hydroxy naphthaldehyde can be obtained in 90% yield (m.p. 80%-81°) starting from β-naphthol.
3. Riemer-Tiemann reaction of thiophone, pyrrole and indoles, gives thiphene 2-aldehyde, pyrrole-2-aldehyde and indole-3-aldehyde respectively.

5.15 2,4,6- TRIMETHYLBENZOIC ACID

It is obtained by saponifiction of methyl 2,4,6-trimethyl benzoate. Normally saponification is carried out with potassium hydroxide. In this case, the main problem arises due to insolubility of potassium hydroxide in organic solvents like toluene. It has been shown that using a crown ether like dicyclohexyl-18-crown-6, the potassium hydroxide is soluble in toluene; this is due to the formation of potassium salt of the crown ether, which is soluble in toluene. Even sterically hindered esters can be saponified by using potassium hydroxide complex in toluene.

H_3C CH_3 O C—OCH_3 CH_3 — K^+ dicydohexyl [18°] crown-6 / Toluene, reflux, 30 min → H_3C CH_3 COOH CH_3

Methyl 2, 4, 6-trimethyl benzoate

2, 4, 6-Trimethyl benzoic acid

Materials

Methyl 2, 4, 6-trimethyl benzoate	5 g
Potassium hydroxide	3 g
Toluene	25 mL
Dicyclohexyl [18] crown-6	0.1 g

Procedure

To a solution of methyl 2,4,6-tmethylbenzoale (5 g) in toluene (25 mL) is added dicyclo hexyl [18] crown-6 (0.1 g) and potassium hydroxide (3 g). The mixture is stirred and refluxed for 45 min. The solution is cooled, and acidified. The separated 2,4,6-trimethyl benzoic acid is crystallised from alcohol in 95% yield. Determine its m.p.

Notes

1. C.J. Pedersen and H.K. Friensdorff, Angew. Chem. Ind. Ed. Engl., 1972, **11**, 16; C.J. Pedersen, J. Am. Chem. Soc., 1967, **89**, 2485; 7017; 1970, **92**, 386, 391.
2. K^+ Dicyclohixnyl [18] crown-6 salt is represented as

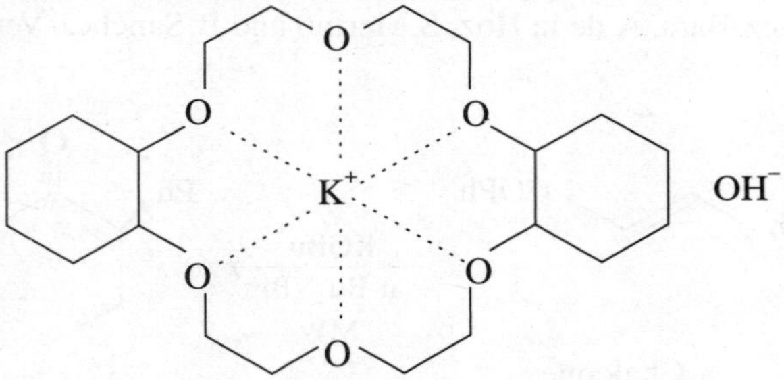

$$K^+ \text{ (dicyclohexyl [18] crown-6 complex, six O atoms coordinated to } K^+) \quad OH^-$$

5.16 MISCELLANEOUS APPLICATION OF PTC

(i) Benzoin condensation Benzoin condensation of aldehydes is catalysed by quaternary ammonium cyanide in a two phase system (J. Soloder, Tetrahedron Lett., 1971, 287)

$$C_6H_5CHO + C_6H_5CHO \xrightarrow{PTC} C_6H_5\,CHOHCO\,C_6H_5$$

Benzoin (acylons)

Benzoins or acyloins can also be obtained by stirring aliphatic or aromatic aldehydes with a quaternary catalyst, N-laurylthiazolium bromide in aqueous phosphate buffer at room temperature (W. Tagaki and H.Hara, J. Chem. Soc. Chem. Commun., 1973, 891). This is a type of acyloin condensation.

Benzoin condensation can also be carried out with either aqueous KCN/neat aromatic aldehyde or solid KCN/aldhyde dissolved in benzene or acetonitrile at 25-60° using 18-crown-6 as catalyst.

(ii) Darzen's Reaction The condensation of aldehydes and ketones with chloroacetonitrile in presence of aqueous sodium hydroxide using a PTC (benzyl triethyl ammonium cloride) gives the corresponding epoxides (A. Jonczyk, M. Fed orynski and M. Makosa, Tetrahedron Lett., 1972, 2395)

$$\underset{R'}{\overset{R}{>}}C{=}O + ClCH_2CN + \underset{aq.}{NaOH} \xrightarrow{C_6H_5CH_2\overset{+}{N}Et_3\overset{-}{Cl}} \underset{R'}{\overset{R}{>}}\underset{O}{C{-}}CH_2{-}CN$$

See also preparation of 1-oxaspiro [2,5]-octane-2-carbnitrile (Section 5.11)

In the above condensation use of an acedic ketone like phenylacetone result in alkylation

$$\underset{\text{Phenyl acetone}}{C_6H_5\ CH_2COCH_3} + ClCH_2CN + \underset{aq}{NaOH} \xrightarrow{Q^+X^-} C_6H_5{-}\underset{NCCH_2}{CHCOCH_3}$$

(iii) Michael Reaction The michael reaction of 2-phenyl-cyclonexanone with chalcone can be carried out under PTC conditions to give 2,6-disubstituted cyclohexanone derivative in high disteroselectivity (E. Diez-Bara, A.de la Hoz, S.Merino and P. Sanchez-Verdu, Tetrahectron Lett., 1997, **38**, 2359).

2-Phenyl cyclohexanone + Chalcone (Ph–CH=CH–COPh) $\xrightarrow[\text{n-Bu}_4\text{NBr, MW}]{\text{KOBu}^+}$ product (99% ee)

Another example of solid state Michael addition reaction is the reaction of nitromethane to chalcone in presence of alumina under MW irradiation to give the adduct in 90% yield (A. Boruah, M. Boruah, B. Prajapati and J.S. Sandhu, Chem. Lett, 1997, 965)

Chalcone + CH_3NO_2 (Nitro methane) $\xrightarrow[\text{Alumina}]{\text{MW}}$ Adduct (PhCOCH$_2$CH(CH$_2$NO$_2$)Ph) 90%

(iv) Welliamson Ether synthesis Alcohols on reaction with alkyl halide in presence of NaOH solution and a PTC give ethers in excellent yield (H.H. Freeman and R.A. Dubois, Tetraheron Lett., 1975, 3251)

$$C_7H_{17}OH + C_4H_9Cl \xrightarrow[\text{NaOH soln.}]{\text{PTC}} C_8H_{17}OC_4H_9 + \underset{\text{Byproduct}}{C_8H_{17}OC_8H_{17}}$$

PTC used is tetrabutylammonium bisulfate.

(v) Witting Reaction In the PTC catalysed Witting reaction, the PTC viz., alkyltriphenylphosphonium salt is reacted with aqueous NaOH to generate the ylide, which in turn combines with aldehydes to produce olefins (W. Tagaki, I. Inouse, Y. Yano and T. Okonogi, Tetrahedron Lett., 1974, 2587; S. Hung and I. Stemmler, Tetrahedron Lett., 1974, 3151).

$$(C_6H_5)_3\overset{+}{P}-CH_2C_6H_5\overset{-}{Cl} + \underset{\text{(aq)}}{NaOH} \xrightarrow{CH_2Cl_2} \underset{\text{ylide}}{\left[(C_6H_5)_3P=CHC_6H_5\right]} \xrightarrow{RCHO} \underset{\text{Alkene}}{RCH=CHC_6H_5} + (C_6H_5)_3PO$$

(vi) Wittig-Horner Reaction This is a modification of witting reaction and can be performed under PTC conditions. In the **PTC catalysed Wittig-Horner Reaction**, the phosphine oxide, is reacted with a ketone in NaOH solution using a PTC, tetraalkylammunium salt or a crown ether as a catalyst (M. Mikolajczyk, S.Grzejszczyk, W. Midura and A. Zatorki, Synthesis, 1976, 396)

$$Et_2OP(O)CH_2R + \begin{matrix}R' \\ R''\end{matrix}\!\!>C=O + \underset{\text{aq.}}{NaOH} \xrightarrow{PTC} RCH=C\!\!<\!\!\begin{matrix}R' \\ R''\end{matrix} + (EtO)_2PO_2Na$$

$R = CN, CO_2Et, SC_6H_5$ etc.

(vii) 3-Aryl-2H-1, 4-benzoxazines These benzoxazines, known for their anti-inflammatory activity are prepared by the reaction of 2-aminophenols with phenacyl bromide in presence of a PTC (G. Sabita and A.V. Subba Rao, Synth. Commun., 1987, **17**, 341)

$$\underset{\text{o-Aminophenols}}{R\text{-}C_6H_3(OH)(NH_2)} + \underset{\text{Phenacyl bromide}}{C_6H_5COCH_2Br} \xrightarrow[\text{aq.}K_2CO_3]{\text{PTC}} \text{3-Phenyl-2H-1,4-benzoxazines}$$

(viii) 2-Aroylbenzofurans These are obtained by PTC catalysed reaction of an o-hydroxyacetophenone with phenacyl bromide in aqueous K_2CO_3 solution (G. Sabitha and A.V. Subba Rao, Synth. Commun., 1987, **17**, 341)

$$\text{R-(o-hydroxyacetophenone: OH, COCH}_3\text{)} + C_6H_5COCH_2Br \xrightarrow[\text{aq.}K_2CO_3]{\text{PTC}} \text{R-(benzofuran with COPh, CH}_3\text{)}$$

o-Hydroxy acetophenanes — Phenacyl bromide — 2-Aroyl benzofurans

PTC = tetrabutlyammonium hydrogen sulphate

(ix) 1, 4-Benzoxazines These are obtained by the reaction of an N-benzoyl o-aminophenol with 1,2-dibromoethane using solid sodium hydroxide and mixture of acetonitrite and methylene chloride as solvent in the presence of PTC (aliquat) (G. Coudert, G. Guillaumet and B. Loubinous, Synthesis, 1979, 541)

$$\text{R}^1,\text{R}^2\text{-C}_6H_2(NHCOC_6H_5)(OH) + Br(CH_2)Br \xrightarrow[\substack{\text{Solid NaOH} \\ CH_3CN - CH_2Cl_2}]{\substack{(C_8H_{17})_3N^+CH_3Cl^- \\ \text{(aliquat)}}} \text{R}^1,\text{R}^2\text{-benzoxazine (N-COC}_6H_5\text{)}$$

(x) oxidation

(a) Potassium premanganate is best used as an oxidant in presence of a phase transfer catalyst like tetraalkyl ammonium or phosphonium salts. Some examples are given below.

$$\underset{\text{Toluene}}{C_6H_5CH_3} + \underset{\text{aq}}{KMnO_4} \xrightarrow{\substack{\text{Cetyltrimethyl} \\ \text{ammonium chloride}}} \underset{\text{Benzoic acid}}{C_6H_5COOH}$$

In place of PTC, crown ether can also be used.

$$\underset{\text{1-Decene}}{C_7H_{15}CH_2CH = CH_2} + \underset{\text{aq}}{KMnO_4} \xrightarrow[\text{R.T.}]{C_{16}H_{33}N^+(CH_3)_2CH_2C_6H_5\bar{Cl}} \underset{77\%}{C_7H_{15}CH_2COOH} + \underset{8\%}{C_7H_{15}COOH}$$

$$\text{cyclooctene} + KMnO_4 \xrightarrow[\text{NaOH}]{C_6H_5CH_2\overset{+}{N}Et_3\overset{-}{Cl}} \text{cyclooctane-1,2-diol (H, OH, H, OH)}$$

Cis-1,2-cyclooctane
55%

$$\underset{\text{Stilbene}}{C_6H_5CH{=}CH_2} + \underset{\text{aq.}}{KMnO_4} \xrightarrow[\text{2-3 hr stirring}]{(CH_3)_4\overset{+}{N}\overset{-}{Br}} \underset{\substack{\text{Benzoic acid} \\ 95\%}}{C_6H_5COOH}$$

(b) Chromate oxidations are generally effected by using commercially available quaternary salt ADOGEN 464 [a mixture of trialkyl (C_6-C_{10}), methylammonium chloride]. Using this PTC, alcohols like benzyl alcohol, 1-phenylethanol, cinnamyl alcohol and cyclodecanol give good yields of the corresponding aldehydes (G.A. Lee and H.H. freedman, Tetrahedron Lett,. 1976, 1641)

An commercial resin, Amberlyst A-26 can be converted into $HCrO_4^-$ form by stirring CrO_3 (5 g) in water (35 mL). The resin so obtained (1.3 mol CrO_3 per g of resin) is stable at room temp. for several weeks. This has been used for the oxidation of allylic or benzylic alcols in varous refluxing solvents (like benzene, chloroform, THF) to give good yields of the corresponding aldehydes. In this oxidation, carboxylic acids are not obtained (G. Cainelli, G. Cardillo, M. Orena and S. Sandri J. Am. Chem. Soc., 1976, **98**, 6737; G. Cardillo G. Giamecian and S. Sandri, Tetrahedron Lett., 1976, 3985).

(c) Hypochlorite oxidation can be conveniently effected by quaternary cations. Some example are give below.

$$\underset{\text{Benzyl alcohol}}{C_6H_5CH_2OH} + \underset{\text{aq.}}{NaOCl} \xrightarrow[\text{CH}_2\text{Cl}_2,\ 1.5\text{ hr}]{\text{Bu}_4\text{N}^+\text{X}^-} \underset{76\%}{C_6H_5CHO}$$

$$\underset{\text{Aliphatic alcohol}}{RCH_2OH} + NaOCl \xrightarrow[\text{Slow reaction}]{\text{Bu}_4\text{N}^+\text{X}^-} (RCHO) \rightarrow RCO_2H$$

$$\text{Cycloheptanol} + NaOCl \xrightarrow[\text{EtO Ac},\ 2\text{ hr}]{\text{Bu}_4\text{N}^+\text{X}^-} \text{Cycloheptanone (89\%)}$$

$$\underset{\text{Octylamine}}{C_7H_{15}CH_2NH_2} + NaOCl \xrightarrow[\text{EtO Ac},\ 30\text{ min}]{\text{Bu}_4\text{N}^+\text{X}^-} \underset{\text{Cyanoheptane (60\%)}}{C_7H_{15}CN}$$

(d) Osmium tetroxide (or Ruthenium Tetroxide). Peracids are used for oxidation of alkenes in presence of quaternary ammonium salts or trialkylammines (Charles M. Starks and Charles Liotta, Phase Transfer catalysis. Principles and techniques, Academic Press, Inc. Ny, P. 310)

(e) Air oxidations is effected in case of fluorene in presence of tricaprylmethylammonium chloride in benzene and NaOH solution.

$$\underset{\text{Fluorene}}{\text{[fluorene]}} + O_2 \xrightarrow[\text{aq NaOH, benzene}]{\text{Quaternary salt}} \underset{70\%}{\text{[fluorenone]}} + H_2O$$

(f) Peracid Oxidations PTC has also been used for peracid oxidation (J.P. Adamson, R. Bywood, D.T. Eastlick, Y. Gallagler, D. Walker and E.M. Wilson, J.Chem. Soc., Perkin Trans, 1975, 2050). One example is given below:

$$\underset{\text{in } CH_2Cl_2}{(C_6H_5)C{=}NNH_2} + CH_3CO_3H + \underset{\text{aq}}{NaOH} \xrightarrow[\substack{I_2\ (\text{Cocatalyst}) \\ 0°,\ 45\ \text{min}}]{(C_8H_{17})_3N^+C_3H_7Cl^-} \underset{95\%}{(C_6H_5)_2C{=}N{=}N}$$

Conclusion

Procedures using phase transfer catalysts have gained wide applications in organic synthesis. It offers very convenient conditions for a variety of reaction (transformations). Though most of such reactions were carried out under traditional conditions, but the use of PTC increases the yields and purity of the products. In fact, this technique provides much simpler procedures for the oxidation reactions and also for the isolation of products. There are many reactions that do not proceed satisfactorily unless carried out under PTC conditions.

The procedures described include esterification, saponification and oxidation. Besides these, some reactions like Baker-Venkataraman rearrangement, Darzen condensation, Hoffmann carbylamine, reaction, Perkin reaction, Reimar-Tiemann reaction and Williamson's ether synthyis have also been described.

Chapter 6

Transformations Using Sonication

6.1 BENZYL CYANIDE

It is prepared by the reaction of benzyl bromide in toluene with potassium cyanide. The reaction is calatylised by alumina on sonication to give the substitution products vis. benzyl cyanide in 76% yield (T. Ando, S. Sumi, T. Kawate, J. Ichihara and T. Haafusa, J. chem. Soc. Chem. Commus, 1984, 439)

$$C_6H_5CH_2Br + C_6H_5CH_3 + KCN + Al_2O_3 \xrightarrow{))))} C_6H_5CH_2CN$$

Benzyl bromide — Benzyl cyanide (76%) B.P. 161.8°

Notes

1. In the above procedure, the formation of Friedel-crafts alkylation product is not observed. Without the use of ultrascund 83% of the Friedel-crafts product $C_6H_5\text{-}CH_2\text{-}C_6H_4\text{-}CH_3$(p) is obtained on stirring. A possible explanation involves ultrasonic dispersion of KCN on the alumina surface, with deceases the Friedel-Crafts activity while promoting the nucleophilic displacement of CN^- on its surface.
2. Benzyl cyanide can also be obtained in good yield by stirring a mixture of benzyl bromide in an organic solvent with aqueous KCN in presence of a PTC (hexadecyl tributyl-phosphonium bromide, $C_{16}H_{33}P^+(C_4H_9)_3Br^-$ at about 100°. This is a general method for the conversion of bromide into cyanide (C.M Starks, J-Am. Chem. Soc. 1971, **93**, 195; N. Sugimoto, T. Fujita, N. Shigematsa and A. Yada, Chem. Bul., 1962, **10**, 427).
3. Use of benzoyl chloride or bromide in place of benzylcyanide give **benzoyl cyanides** (K.E. Koening and W.P. Weber, Tetrahedron Lett, 1974, 2275, H. Habner, Ann. Chem. Pharm., 1861, **120**, 330

$$\underset{org}{C_6H_5\overset{O}{\overset{\|}{C}}—Cl} + \underset{aq}{NaCl} \xrightarrow{B_4N^+X^-} \underset{60\text{-}70\%}{C_6H_5\overset{O}{\overset{\|}{C}}—CN} + NaCl$$

Alternatively, benzoyl cyanide can also be obtained from benzoyl chloride by treatment with KCN in acetonitrile at 50° and subjecting it to sonication even in the absence of PTC (T. And, J. Kawate and H. Hanatusa, Synthesis, 1983, 637)

$$C_6H_5\overset{O}{\overset{\|}{C}}—Cl \xrightarrow[50°,\ ))))]{KCN,\ MeCN} \underset{80\%}{C_6H_5\overset{O}{\overset{\|}{C}}—CN}$$

6.2 BIPHENYL

It is obtained by the homocoupling of bromobenzene in presence of lithium in tetrahydrofuran by subjecting the mixture to sonication[1].

$$\underset{\text{Bromobenzene}}{C_6H_5Br} \xrightarrow[\substack{)))) \\ 10\ min}]{Li,\ THE} \underset{\text{Biphenyl}}{C_6H_5—C_6H_5}$$

Procedure

A mixture of dry pure bromobenzene (0.02 mol), lithium wire (0.02 mol) in THF (20 mL) was immersed in an ultrasonic both (117 W. 50 KHz) for 10 min. The reaction mixture is treated with few drops of ethanol (to destroy any lithium left) and extracted with ether. Distillation of ether gave biphenyl, in 80 of yield. It crystallised from alcohol, m.p 70°.

Notes

1. B.H. Han and P. Boudjouk, Tetraheolron left, 1981, **22**, 2757; T.D. Lash and D. Berry, J. Chim. Edu., 1985, **62**, 85
2. Coupling of benzyl halides in presence of copper or nickel powder generated by lithium reduction of the corresponding halide in the presence of ultrasound gave high yields of dibenzyl, m.p 53° (P. Boudjouk, W.H. Thompson, W.H. Ohbrom and B.H. Han, Organometallics, 1986, **5**, 1257). The reaction is know as **Ullinann's Reaction.**

$$CuBr_2 \xrightarrow[))))]{Li,\ THF} Cu^* \xrightarrow[))))]{C_6H_5CH_2Cl} C_6H_5-CH_2-CH_2-C_6H_5$$

80%
Dibenzyl

3. Ulimann coupling of activated aryl halides in DMF with high intensity ultrasound gave much better yield. Thus, o-iodonitrobenzene on sonication in presence of Cu, DMF at 60° gave good yield of 2, 2′-dinitro biphenyl, M.P. 124-125°. (L. Lindley, J.P. Lorimer and T.J. Mason, Ultrasonics, 1986, **24**, 292)

$$\text{o-Iodonitrobenzene} \xrightarrow[))))]{Cu,\ DMF,\ 60°} \text{2,2'-Dinitrobiphenyl (80\%)}$$

4. **Substituted biphenyls** have also been obtained by **aqueous Suzuki reaction** (N.E. Leadbeater and M. Morco, J. Org. Chim., 2003, **68**, 888; L. Bai, J.-X. Wang and Y. Zhang, Green chem., 2003, **5**, 615. The procedure consist in heating aryl halides and phenyl boronic acid in presence of catatic amount of Pd/C in aqueous medium with MW heating (R.K. Arvela and N.E. Leadbeater, Org. lett., 2005, **7**, 210)

$$R-C_6H_4-X + C_6H_5B(OH)_2 \xrightarrow[\text{MW, 150°, 5 min}]{Pd(OAc)_2,\ Na_2CO_3,\ TBAB,\ H_2O} R-C_6H_4-C_6H_5$$

X = Cl, Br, I; R = Me, OMe, COMe — Phenyl boronic acid — Substituted biphenyls 62-91%

Alternatively, **Suzuki Cross-Coupling** of aryl boronic acids with aryl halides in polyethylene glycol (PEG) using microwaves (V.V. Narmboodiri and R.S. Varma, Green Chem, 2001, **3**, 146) gives biphenyls

$$C_6H_5Br + (HO)_2B-C_6H_4-CH_3 \xrightarrow[\text{PEG-400, MW, 50 Sec.}]{Pd\,Cl_2,\ KF} C_6H_5-C_6H_4-CH_3$$

75-80%

Using this procedure a number of substituted biphenyls were prepared. (Section 10.15).

5. A novel method of obtaining biphenyl by oxidative coupling of benzene in water, in presence of a couple of additives has been reported (S. Mukhopadhaya, G. Rothenberg, G. Lando, K. Agbaria, M. Kazanci and G. Sasson, Adv. synth. Catal 2001, **343**, 455.

$$\text{Benzene} \xrightarrow[\substack{Zr(OAc)_4,\ 2\ mol\% \\ Co(OAc)_2,\ 2\ mol\% \\ Mn(OAc)_2,\ 2\ mol\% \\ HOAC\text{-}NaOAc,\ 1\ MPa,\ air \\ Water\ 250\ mol\%}]{\substack{PdCl_2,\ 7\ mol\% \\ acetylacetone,\ 2\ mol\%}} \text{Biphenyl}\quad 96\%$$

6.3 BUTYRALDEHYDE

It is obtained by the reaction of 1-chorobutane with lithium and dimithyl formamide under ultrasonic irradiation. The reaction is known as **Bouveault reaction** (C. Petrier, A.L. Gemal and J.L. Luche, Tetrahedron Lett, 1982, **23**, 3361).

$$\underset{\text{1-Chlorobutane}}{CH_3CH_2CH_2CH_2Cl} \xrightarrow{Li} \underset{\text{N-Butyllithium}}{CH_3CH_2CH_2CH_2Li} \xrightarrow{DMF}$$

$$\left[CH_3CH_2CH_2CH\begin{matrix} OLi \\ NMe_2 \end{matrix} \right] \xrightarrow{H_3O^+} \underset{\text{Butyraldehyde}}{CH_3CH_2CH_2CHO} + HNMe_2$$

Materials

1-Chlorobutane	2.72 g
Lithium	0.15 g
Dimethyl formamide	2.2 g

Procedure

A mixture of 1-Chlorobutane (2.72 g), dry dimethyl formamide (2.2 g) and lithium sand is dry THF (0.1 g in 4 ML. THF) is sonicated in an ultrasound cleaner at 10-20° for 15 min. The reaction mixture is extracted with ether (2 × 30 mL). Ether extract is washed with dil. HCl, Sodium chloride solution, dried (Na_2SO_4) and distilled to give n-butyraldehyde in about 80% yield B.P. 74-75°

Notes

1. This is general method for the conversion of halides into aldehydes
2. Lithium metal is obtained as suspended material in mineral oil. Before use, it is washed with anhydrous THF under inert atmosphere. On sonication it is converted into fine sand of lithium metal.

6.4 CANNIZZARO REACTION

Adehydes which do not have α-hydrogen on treatment with base undergo cannizzaro reaction (self oxidation and reduction) to produce the corresponding alcohol and the salt of the corresponding acid. Thus benzaldehyde gives benzoic acid and benzyl alcohol.

$$\underset{\text{Benzaldehyde}}{C_6H_5CHO} \xrightarrow[\text{)))), 10 min}]{Ba(OH)_2,\ EtOH} \underset{\text{Benzyl alcohol}}{C_6H_5CH_2OH} + \underset{\text{Benzone acid}}{C_6H_5COOH}$$

Procedure[1]

Dissolve fresh distilled benzaldehyde (0.2 mol) in ethanol (15 mol). To the solution add barium hydroxide (12 g). Subject the mixture to low intensity ultrasound (cleaning bath) for 10 min. Extract the mixture with ether (3 × 20 mL) after addition of water (30 mL). Cool the alkaline solution and acidify with acid (HCl). Cool the acidified solution to obtain benzoic and m.p. 121°.

Dry the combined elder extract (obtained above) over anhydrous magnesium sulphate and distil. The residual product is benzyl alcohol, b.p. 205-206°. Total yield 100%

Note

1. A. Fuentes and J.V. Sinisterra, Tetraredron left., 1986, **27**, 2967
2. No reaction takes place in the above case in the absence of ultrasound.
3. Normal cannizzo reaction is conducted by reacting aldehyde with potassium hydroxide solution at rooms temperature. The reaction is worked up as described above. (V.K. Ahluwalia and Renu Aggarwal, comprehensive practical organic chemistry; Preparation and Quantitative analysis, Universities pres 2000. Page 39 and its references cited there in).

6.5 2-CARBETHOXY CYCLOPENTANONE

It is obtained by the **Sonochemical Dieckmann Condensation** of diethyl adipate by reaction with potassium in tetrahydrofuran.

$$\underset{\text{Diethyl adipate}}{EtO_2C(CH_2)_4CO_2Et} \xrightarrow[\text{)))), 5 min}]{\text{K, toluere}}$$

O

CO_2Et

2-Carbethoxy cyclopentanone

Procedure[1]

Diethyl adipate (0.2 mol) in toluene (20 mL) containing potassium (0.4 mol) was sonicated using low intensity ultrasonic bath for 5 min. Excess potassium was decomposed with alcohol (1-2 mL) and the mixture extracted with ether. The ether extract was dried and distilled to yield 2-carbethoxy cyclopentanone in 80% yield.

1. Cited in a review by J.M. Khurana, Chemistry Education, 1990 (Oct, Dec.) P. 27. J.L. Luche, C. Petrces and C. Dupu. Tetrahedron left., 1985, 26, 753
2. On sonication, potassium is easily transformed to a silver blue suspension in toluene. The ultrasonical dispersed potassium is very helpful in Dieckmann condensation.
3. The starting diethyl adipate is obtained by esterification adipic add with ethanol in presence of conc. H_2SO_4

6.6 N-2-CHLOROPHENYL ANTHRANILIC ACID

It is obtained[1] by the **Ullmann condensation** of 2-chlorobenzoic and with 2-chloroaniline in presence of copper powder and cuprous iodide in boiling diethyl formamide using ultrasonic irradiation.

COOH, Cl — 2-Chloro benzoic acid + Cl, H_2N — 2-Chloro aniline → (Cu powder; Cuprous iodide; ultrasonic irradiation; 20 min) → CO_2H Cl, NH — N-2-Chlorophenyl anthranilic acid

Materials

2-Chlorobenzoic acid	2.2 g
2-Chloroaniline	1.5 g
Copper powder	0.15 g
Cuprous iodide	0.15 g
Potassium carbonate	0.8 g

Procedure

A mixture of 2-chlorobenzoic acid (2.2 g) 2-chloroaniline (1.5 g), potassium carbonate (0.8 g) and cuprous iodide (0.15 g) in boiling DMF (8 mL) is heated for 20 min with ultrasonic irradiation. The reaction mixture is pound on to water (30 mL). The separated product is filtered, treated with potassium carbonate solution. The clear filtrate is acidified with dilute acetic acid. The sparated product is crystallised from alcohol, m.p 196-97°. Yield 78%.

1. R. Carrasco, R.F. Pellon, Jose' Elguero, Pillar Goga and Juan Antorio Paez, Synthetic Commum. 1989, **19**, 2077.
2. The procedure using ultrasound is convenient and the reaction is complete in 20 min. compared to 5-6 hs by the usual procedure. The reaction is much cleaner and gives good yield and purity of the product.

6.7 CINNAMALDEHYDE

It is obtained by sonication of cinnanayl alcohol with manganese dioxide in hexane at room temperature for 2.5 hs. in 70% yield (T. Kimura, F. Fujita and A. Ando, Chem. Lett., 1998, 137).

$$\underset{\text{Cinnamyl alcohol}}{C_6H_5CH{=}CH\,CH_2OH} \xrightarrow[\text{)))) RT, 2.5 hr}]{MnO_2/\text{hexane}} \underset{\substack{\text{Cinnamaldehyde} \\ 70\%}}{C_6H_5CH{=}CHCHO}$$

Procedure

A mixture of cinnamyl alcohol, manganese dioxides and hexane is sonicated for 2.5 hr. The mixture is diluted with water and extracted with ether. The ether extract is dried ($MgSO_4$) and distilled to give cinnamaldehyde in 70% yield (b.b. 248°)

1. Cinnamaldehyde is also obtained by the oxidation at cinnanyl alcohol with chromium trioxide in conc. H_2SO_4 or with chromium trioxide in pyridine. (V.K. Ahluwalia Renu Aggarwal, Comprehensive Organic Chemistry; preparation and quantitative analyses. Universities Press, 2004, Page 137 and 206).
2. Using the sonication procedure, 1-phenylethanol or geraniol gives the corresponding aldlhydes In this cases oxidation is best carried out in octane as solvent.

6.8 CYCLOHEXANONE

It is obtained by the oxidation of cyclohexanol with solid $KMnO_4$ in hexane or benzene under sonication in 53% yield (J. Yamakawi, S. Sume, T. Ando and J. Harafusa, Chemistry Lett., 1983, 379)

$$\underset{\text{Cydohexanol}}{\text{(OH-cyclohexane)}} \xrightarrow[\text{RT,))))}]{KMnO_4/\text{hexane or benzene}} \underset{\substack{\text{Cydohexanone} \\ 53\% \\ \text{b.p. 154-155°}}}{\text{(cyclohexanone)}}$$

Procedure

A mixture of cyclohexanol and $KMnO_4$ in hexane is sonicited for 2 hr at room lempentus. The mixture is extracted with ether, and ether distilled to give cyclohexanone in 53% yield.

Notes

1. This is a convenient method for the oxidation of secondary alcohols to ketones. Thus octan-2-ol ($C_6H_{13}CHOHCH_3$) gives 2-octanone ($C_6H_{13}COCH_3$) in 92.8% yield (4 hr sonication)
2. This procedure is more convenient compared to oxidation with $K_2Cr_2O_7/H_2SO_4$ (dil.)

6.9 7.7-DIBROMOBICYCLO (4.1.0) HEPTANE

It is obtained by the addition of dibromo carbene (generated from bromoform and alkali under sunication) to cyclohexene.

$$\text{Cyclohexene} \xrightarrow[\text{)))), 10 min}]{\text{NaOH, CHBr}_3} \text{7,7-Dibromobicyclo (4.1.0) heptane}$$

Procedure

A mixture of cyclohexene (0.2 mol), bromoform (0.2 mol) and sodium hydroxide (powdered, 0.3 mol) is sonicated using low intersity ultrasound (using a cleaning bath) for 10 min. The formed product obtained in 90% yield is isolated by ether extraction and subsequent distillation, b.p. 100° 8 mm.

Notes

1. Using chloroform in place of bromoform, 7.7-dichlorobicyclo (4.1.0) heplane is obtained in 85% yield.
2. Use of sonication and stirring given much better yield.
3. For the generation of dibromo or dichlorocarbene see S.L. Regen and A. Singh, J. Org. Chem., 1982, **47**, 1587
4. It can also be prepared by the reaction of cyclohexene with aqueous NaOH and $CHBr_3$ in presence of tetra-n-butyl ammonium chloride (see Section 5.5)

6.10 ETHYL 2-HYDROXY-2-PHENYL ACETATE

It is obtained by the **Reformatsky reaction** of an aldehycle (benzaldehyde) with α-haloeater (ethyl bromoacetate) in presence of zinc (activated with iodine) using ultra sound.

$$\underset{\text{Benzaldehyde}}{C_6H_5CHO} + \underset{\text{Ethyl bromo acetate}}{BrCH_2CO_2Et} \xrightarrow[\text{)))), 15 min}]{\text{Zn/I}_2\text{, dioxane}} \underset{\substack{95\% \\ \text{Ethyl 2-hydroxy-2-phenylacetate}}}{C_6H_5\overset{\overset{\displaystyle OH}{|}}{C}HCH_2CO_2Et}$$

Procedure[1]

Benzaldehyde (0.02 mol) and ethyl bromoacetate (0.02 mol) was dissolved in dioxane (20 mL). To the above solution zinc (5 g) and iodine (0.1-0.2 g) added and the mixture subjected to low intensity ultrasound (using a cleaning bath) for 15-20 min. The progress of the reaction was monitored by TLC. When the reaction was complete,water (30 ML) added and extracted with ether. Drying the ether extract with anhydrous magnesium sulphate and subsequent distillation yielded ethyl 2-hydroxy-2-phenyl acetate in 98% gild, B.P. 105°/0.3 mm.

Notes

1. B. H. Han and P. Boudjouk, J. Org. Chem., 1982, **47**, 5030
2. Using the above procedure, the reaction of ethyl bromo acetate with benzophenone give ethyl 3,3-diphenyl 3-hydroxypropionate, m.p. 87-88°.

$$BrCH_2\overset{\overset{O}{||}}{C}-OC_2H_5 + (C_6H_5)_2CO \xrightarrow[\text{)))), 20 min}]{Zn/I_2,\ \text{dioxane}} C_6H_5-\underset{\underset{C_6H_6}{|}}{\overset{\overset{OH}{|}}{C}}-CH_2-\underset{\underset{O}{||}}{C}-OC_2H_5$$

Ethyl 3,3-diphenyl
3-hydroxy propionate

6.11 ETHYL PHENYL ETHER (PHENETOLE)

It is obtained by the reaction of phenol and ethyl iodide in polyethylene glycol (PEG) in presence of potassium hydroxide using sonication.

$$\underset{\text{Phenol}}{C_6H_5OH} + \underset{\text{Ethyl iodide}}{C_2H_5I} \xrightarrow[\text{))))}]{KOH,\ PEG} \underset{\text{Ethyl phenyl ether (80\%)}}{C_6H_5OC_2H_5}$$

Procedure

A solution of phenol (0.02 mol), ethyl iodide (0.02 mol) and powdered potassium hydroxide (3 g) in polyethylene glycol (PEG, 20 mL) was subjected to sonication using low intensity ultrasound (cleaning bath). Working up of the reaction mixture gave 80% yield of ethylphenyl ether (phenetole) b.p. 172°

Notes

1. R.S. Davidson, A. Safdar, J.D. Spencer and D.W. Lewis, Ultra sonics, 1987, **25**, 35
2. Using this method a numbers of mixed ethers could by prepared by using substituted phenols and different alkyl halides.
3. This method can also be used for synthesising esters by using a carboxylic and in place of phenol.

$$RCOOH + R'X \xrightarrow[\text{))))}]{KOH,\ PEG} \underset{\text{esters}}{RCO_2R'}$$

6.12 MISCELLANEOUS APPLICATIONS OF ULTRASOUND

(i) Hydrosis of Nitriles

Hydrosis of nitriles to carboxylic acids under basic condition is improved by sonication (J. Elguero, J. Goga, L. Lissavetzky and A.M. Valdeomillos, C.R. Acad. Sci. Paris, 1984, 298, 877). Thus benzylcyanide on hydrolysis give phenyl acetic acid

$$\underset{\text{Benzylcyanide}}{C_6H_5CH_2CN} \xrightarrow[))))]{HO^-\ H_2O} \underset{\text{Phenyl acetic acid}}{C_6H_5CH_2COOH}$$

(ii) Solvolysis

Solvolysis of tert.butyl chloride in aqueous alcohol gives tertary butyl alkyl ether. The rate is enhanced about two times using sonication. However, use of a probe generator enhances the rate twenty times at 10° (J.P. Lorimer, T.M. Mason and B.P. Mistry. Ultrasonics, 1987, **25**, 23)

$$\underset{\text{Tertiary butylchloride}}{(CH_3)_3C{-}Cl} \xrightarrow[))))]{ROH/H_2O} \left[(CH_3)_3C^+\ Cl^-\right] \longrightarrow \underset{\text{Tert. butyl alkyl ether}}{(CH_3)_3C{-}OR} + HCl$$

(iii) Formation of Azides and Sulphocyanides

Primary alkyl halides on treatment with aqueous sodium azide give the corresponding alkyl azides using ultrasound (H. Priebe, Acta. Chem. Scand. Ser. B. 1984, **38**, 895)

$$RX \xrightarrow[))))]{NaN_3/H_2O,\ 60°} RN_3 \qquad \begin{matrix} R = \text{alkyl yield } 20\% \\ R = \text{allyl yield } 86\% \end{matrix}$$

Sulphocyanides are obtained from alkyl bromidies on sonication with KCNS in presence of a quaternary ammonium salt. Sonication gives 62% yield compound to 43% under usual procedure.

$$CH_3CH_2CH_2CH_2Br \xrightarrow[))))]{KSCN/H_2O/Bu_4\overset{+}{N}\overset{-}{Br},\ RT,\ 6\ hr} \underset{62\%}{CH_3CH_2CH_2CH_2SCN}$$

(iv) Cyclopropanation

In this procedure, sonochemically activated zinc and methylene iodide is used. The yield is 91% compared to about 50% by the usual route.

$$Me(CH_2)_7CH{=}CH(CH_2)_7CO_2Me \xrightarrow[))))]{Zn,\ CH_2I_2} Me(CH_2)_7\text{-cyclopropane-}(CH_2)_7CO_2H$$

This procedure can also be scaled up (O. Repic, P.Y. Lee and N. Giger, Org. Prepn Proc. Int., 1984, **16**, 25; H. Tso, T. Chou and S. Hung, J. Chem. Soc. Chem. Commun, 1987, 1552)

(v) Methylenation

Methylenation of carbonyl group normally requires complex reagents. Howiver, it can be easily accomplished by the **Simmon-Smith reagent** (Zn/CH_2I_2) using sonication (J. Yamashita, Y. Inou, T. Kando and H. Hashimolo, Bull. Chem. Soc. Japan, 1984, 57, 2335)

$$RR'C{=}O \xrightarrow[RT,\ ))))]{CH_2I_2/Zn/THF} RR'C{=}CH_2$$

The reaction is mostly applicable to aldehydes and not to ketones. In case of benzaldehyde the yield is about 70% in 20 min.

(vi) N-Alkylation of Amines

N-alkylation of secondary amines (e.g., benzopyrrole) with alkyl iodide (e.g. CH_3I) in presence of KOH and a phase transfer reagent (PEG methyl ether) gives 65% yield of the N-alkylated product (N-methylbenzopyrrole) on sonication compared to about 60% yield under normal conditions (20°, 5 hr).

$$\text{Benzo pyrrole} \xrightarrow[\text{PEG methyl ether,)))), 20°, 30 min}]{\text{MeI/solid KOH/tolulene}} \text{N-methylbenzopyrrole (65\%)}$$

In a similar way, sonication of diphenyl amine with benzyl bromide gives the corresponding N-benzylated product (in 98% yield) compared to 48-70% in 48 hr under reflux conditions (R.S. Davidson, A.M. Patel, A. Safdar and D. Thornthwaite, Tetrahedron Lett., 1983, **24**, 5907).

$$\underset{\text{Diphenyl amine}}{Ph_2\,NH} \xrightarrow[\text{PEG methyl ether, 20°, 1 hr,))))}]{PhCH_2Br/\text{solidKOH/toluene}} \underset{98\%}{Ph_2\,NCH_2Ph}$$

In the absence of PTC, no reaction takes place under sonication.

(vii) Ethers

Ethers are obtained in 80% yield by sonication of alcohol and aryl halides (R.S. Davidson, A. Safdar, J.D. Spencer and D.W. Lewis, Ultrasonics, 1987, 25, 35).

$$\underset{\text{Ethyl alcohol}}{C_2H_5OH} + \underset{\text{Phenyl halide}}{C_6H_5X} \xrightarrow[))))]{\text{KOH, PEG}} \underset{\substack{\text{Ethyl phenyl ether}\\80\%}}{C_2H_5OC_6H_5}$$

(viii) Oxidation

Alcohols are oxidised by solid $KMnO_4$ in hexane and benzene by sonication in an ultrasonic bath (J.M. Khurana and P.K. Sahoo, Syn. Commun. 1992, 1691)

$$RR'CHOH \xrightarrow[\text{RT,))))}]{KMnO_4\text{/hexane or benzene}} RR'C{=}O$$

$R{=}C_6H_{13}$; $R'{=}H$ 92.5%

Similarly, cyclohexanol gives 55% yield of cyclohexanone by oxidation under sonication compared to 4% under usual conditions. In a similar way cinnamic acid is obtained in 82% yield by the oxidation of cinnamaldehyde compared to 4% under usual conditions.

$$\underset{\text{Cyclohexane}}{\text{Cyclohexanol (OH)}} \xrightarrow[\text{RT,))), 5 hr}]{KMnO_4\text{/hexane}} \underset{\substack{\text{Cyclohexanone}\\82\%}}{\text{Cyclohexanone (=O)}}$$

$$\underset{\text{Cinnamaldehyde}}{C_6H_5CH = CH\ CHO} \xrightarrow[\text{RT,))), 3 hr}]{KMnO_4\text{/hexane}} \underset{\substack{\text{Cinnamic acid}\\82\%}}{C_6H_5CH = CH\ COOH}$$

Mangane dioxide, a low relativity oxidant is activated upon sonification (J. Yamakawi, S. Sumi, T. Ando and J. Hanafusa, Chemistry lett, 1983, 379). It is a good oxidising agent for the oxidation of cinnamyl alcohol, geraniol or 1-phenylethanol to the corresponding aldehydes (upon sonication).

$$\text{Cinnamyl alcohol} \xrightarrow[)))]{MnO_2\text{, octane}} \text{Cinnamaledehyde}$$

$$\text{Geraniol} \xrightarrow[))))]{MnO_2\text{, octane}} \text{Gerinal}$$

$$\text{1-Phenylethanol} \xrightarrow[))))]{MnO_2\text{, octane}} \text{Phenylacetaldehyde}$$

Aqueous sodium hypochloride has been used for the oxidation benzylic halides on sonication (J.M. Khurana, P.K. Sahoo, S.S. Titus and G. C. Maikap, Synth. Commun, 1990, 1357)

$$\text{ArRCHX} + \text{NaOCl} \xrightarrow[\text{))))}]{\text{CH}_3\text{CN, RT}} \text{ArRCHOCl} \longrightarrow \text{ArRCHO}$$

(ix) Reduction

The reactivity of the catalysts like platinum, palladium and rhodium black is considerably increased on sonication (A.W. Maltsev, Russ. J. Phys, Chem., 1976, **50**, 995). Thus, hydrogenation of various alkenes can be effected by Pd/C in presence of formic acid using a low intersity ultrasonic fields (cleaning bath, 50 KHz) (J. Jurczak and R. Ostaszewki, Tetrahedron Lett., 1988, 29, 959).

$$\xrightarrow[\text{))))}]{\text{Pd/C, HCO}_2\text{H, 20°}}$$

Alkenes can also be reduced by hydrazine/Pd-C in ethanol at room temperature using an ultrasonic bath (D.H. Shin and B.H. Han, Bull. Korean Chem. Soc., 1985, **6**, 247). This procedure is commercially usefal for the hydrogenation of soyabean oil (K.J. Moulton, S. Koritala and E.N. Frankel, J. Am. Oil Chem. Soc., 1983, **60**, 1257); in this process there is considerable advantage over the usual batch methods, which require much longer time for hydrogenation.

Sonication has also been used for hydrogenolysis of benzyl ethers with H_2/Pd-C in methanol in presence of acetic acid (K.J. Moulton, S. Koritala and E.N. Frankel, J. Am. Oil Chem. Soc, 1983, 60, 125)

$$\text{OCH}_2\text{C}_6\text{H}_5 \xrightarrow[\text{MeOH, CH}_3\text{COOH,))))}]{\text{H}_2\text{/Pd-C}} \text{OH}$$

The activation of nickel powder is also increased by sonication; this in turn is used for the reduction of alkenes (K.S. Suslick and D.J. Casadonte, J. Am. Chem. Soc., 1987, **109**, 3459). The catalytically active nickel is obtained by sonochemical reduction of the salt such as chloride with zinc powder. Under there conditions, the excess of metallic zinc is activated and reduces water present in the medium producing hydrogen gas (C. Petrier and J.L. Luchi, Tetrahedron Lett., 1989, 28, 2347). In this procedure, not only the catalyst but also the reagent (hydrogen) is produced in situ with maximum efficiency and safety. The carbon-carbon double bonds in α, β-unsaturated carbonyl compounds are reduced much faster than the carbonyl group (C. Petrier and J.L. Luchi, Tetrahedron Left., 1987, **28**, 2351). A typical example is given below.

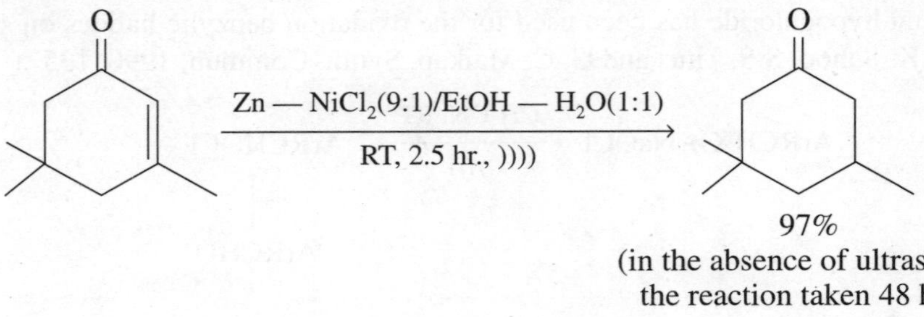

O

$\xrightarrow[\text{RT, 2.5 hr.,))))}]{\text{Zn — NiCl}_2\text{(9:1)/EtOH — H}_2\text{O(1:1)}}$

O

97%
(in the absence of ultrasound,
the reaction taken 48 hrs.)

The selective reduction of C = C in preference C = O depends on the pH of the medium. A typical example in given below

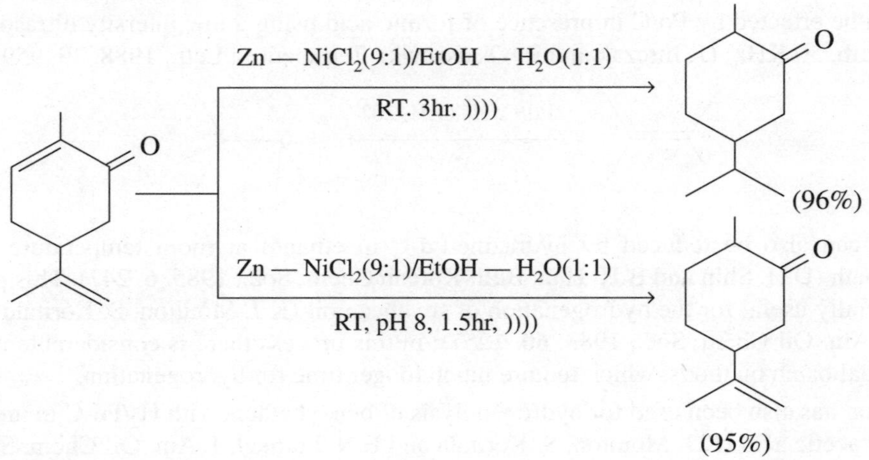

Zn — $NiCl_2$(9:1)/EtOH — H_2O(1:1)
RT. 3hr.))))

O

(96%)

O

Zn — $NiCl_2$(9:1)/EtOH — H_2O(1:1)
RT, pH 8, 1.5hr,))))

(95%)

(x) Hydroboration

It is enhanced considerably by low intensity ultrasound, especially in heterogeneous system. Thus tricyclohexylborane is obtained in 60 min. by the reaction of cyclohexere with BH_3, SMe_2; the traditional proceduol requires 24 hr at room temperature

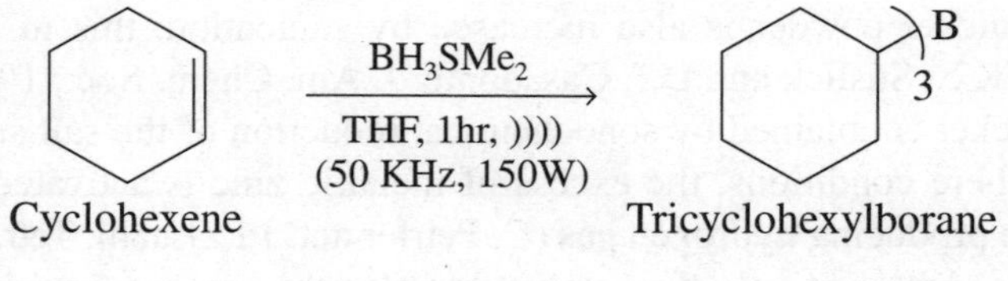

BH_3SMe_2
THF, 1hr,))))
(50 KHz, 150W)

Cyclohexene → Tricyclohexylborane (B, 3)

(xi) Coupling of Aryl Halides

Sonication of aryl, alkyl or vinyl halides with lithium in THF (ultrasonic bath, 117 W, 50 kHz) is a converient procedure for the preparation of biaryls.

$$C_6H_5Br \xrightarrow{\text{Li, THF,)))}} C_6H_5—C_6H_5$$

$$C_6H_5CH_2Cl + Cu^* \xrightarrow[\text{))))}]{\text{Li}} C_6H_5CH_2CH_2C_6H_5$$

$$CuBr_2 \xrightarrow{\text{))))}} Cu^*$$

(xii) Dichloro Carbene

It is generated in situ by reaction between powdered sodium hydroxide and chloroform by sonication. This procedure is simple and avoids the use of a PTC catalyst. Thus sonication of styrene under the above condition gives the adduct in 96% yield. The yield is only 31% (16 hr. Stirring only) (S.L. Regen and A. Singh, J. Org. Chem., 1982, 47, 1587) without sonication.

$$C_6H_5\,CH{=}CH_2 \xrightarrow[\text{)))), stirring}]{\text{NaOH, CHCl}_3} C_6H_5\,CH—CH$$

Cl Cl

96%

(xiii) Conversion of Aryl Halides into the Corresponding Aldehydes

This conversion is conveniently effected by **Sonochemical Bouveault Reaction**. The procedure consists in sonication of aryl halide with lithium in presence of dimethylformamide followed by treatment with dilute acid (C. Petrier, A.L. Gemal and J.L. Luche, Tetrahectron Lett, 1982, **23**, 3361).

$$\underset{\text{Aryl halide}}{RX} + Li \xrightarrow[\text{))))}]{} [\overset{-}{R}\overset{+}{Li}] \xrightarrow{\text{HC(O)NMe}_2} \left[RCH \begin{matrix} \overset{-}{O}\overset{+}{Li} \\ NMe_{2a} \end{matrix} \right] \xrightarrow{H_3\overset{+}{O}} RCHO + Me_2NH$$

Using the above procedure, o-substituted aldehydes can be conveniently prepared.

$$\underset{\text{Aryl halide}}{C_6H_5X} + DMF \xrightarrow[\text{2) n-RBr,)))), 30 min}]{\text{1) Li,)))), 15 min}} \underset{70\%}{o\text{-}R\text{-}C_6H_4CHO}$$

R = alikyl

(xiv) Cannizzaro Reaction

The reaction of aromatic aldehydes with barium hydroxide (catalyst) in presence of ethanol (on sonication) gives the corresponding alcohol in good yield (A. Fuentes and J.V. Sinisterra, Tetrahedron Lett., 1986, **27**, 2967)

$$\underset{\text{Benzaldehyde}}{C_6H_5CHO} \xrightarrow[\text{)))), 10 min}]{Ba(OH)_2,\ EtOH} \underset{\substack{98\text{-}99\% \\ \text{Benzyl alcohol}}}{C_6H_5CH_2OH} + \underset{\substack{1\text{-}2\% \\ \text{Benzoic acid}}}{C_6H_5COOH}$$

(xi) Strecker Synthesis

Amino nitriles can be conveniently prepared by **sonochemical strecker synthesis**. The procedure involves sonication of a mixture of aldehyde or ketone with an amine in presence of KCN and acetic acid (J. Menedez, G.G. Trigo and M.M. Solhuber, Tetrahedron Lett, 1986, 27, 3285)

$$\underset{\text{Aldehyde or Ketone}}{RR'C{=}O} \xrightarrow[\text{))))}]{R^2NH_2,\ KCN,\ AcOH} \underset{\text{Amino nitriles}}{RR'C(CN)(NHR^2)}$$

In a modification of the above process, the reagents are absorbed on the surface of the catalyst before the reaction. Thus, in fact, is a combination of the supported reagent with sonochemical activation; in this procedure the side reactions are supressed (T. Hanafusa, J. Ichihara and T. Ashida, Chemistry Lett, 1987, 687)

$$RCHO \xrightarrow[50°,\ 5\text{-}48\ \text{hrs,))))}]{KCN/Al_2O_3/CH_3CN/NH_4Cl} \underset{80\text{-}100\%}{R{-}CH(CN)(NH_2)}$$

(xvi) Reformatsky Reaction

β-Hydroxyesters are obtained by **Sonochemical Reformatsky Reaction**. The procedure involves the reaction of an aromatic aldehyde with ethylbromoactate, zinc and iodine (catalytic amount) under sonication. The β-hyoboxyester is obtained in 95% yield.

$$RCHO + BrCH_2CO_2Et \xrightarrow[\substack{\text{dioxane} \\ \text{RT, 5 min,))))}}]{Zn/I_2} R\,\underset{\substack{| \\ OH}}{CH}{-}CH_2CO_2Et$$

(xvii) Barbier Reaction

The reaction of ketones with organometallic reagents (generated in situ by the reaction of alkyl halide with Li/THF/sonication) gives the corresponding tertiary alcohol. This reaction is known as **Sonochemical Brabier Reaction**.

This reaction can also be used with unreactive halides. Even reactive halides, e.g., allyl or benzyl halides which normally give Wurtz coupling can also be used (J.L. Luche and J.C. Darmiano, J. Am. Chem. Soc., 1981, **102**, 7926).

$$R^1R^2C{=}O \xrightarrow[\text{RT, 10-15 min))))}]{R^3X/\text{Li/THF}} R^1R^2R^3C{-}OH \quad (70\text{-}100\%)$$

(xviii) Curtius Rearrangement

The rearrangement of an acid azide in non aqueous solvents (like chloroform, benzene or ether) to an isocyanate is known as curtius rearrangement. Normally this reaction is performed under thermal conditions (T. Curtius, J. Prakt. Chem., 1894, **50**, 275; J.H. Saunders and R. Slocombe Chem. Rev., 1948, **43**, 203). However, the yields are low. Under sonochumical conditions the rate of formation is much higher (C.W. Porter and L. Young, J. Am. Chem. Soc., 1938, **60**, 1497)

$$\underset{\text{Benzoyl azide}}{C_6H_5C(=O){-}\ddot{N}{-}\overset{+}{N}{\equiv}N} \xrightarrow{-N_2} \xrightarrow[\text{))))}]{\text{benzene, RT}} \underset{\text{Phenyl isolyanate}}{Ph{-}N{=}C{=}O} + N_2$$

The sonochemical curtius rearrangement can be performed starting with acylchloride as shown below

$$\underset{\text{Acyl chloride}}{R{-}C(=O){-}Cl} + NaN_3 \xrightarrow[\text{)))) RT}]{\text{benzene}} \left[R{-}C(=O){-}N_3\right] \longrightarrow RN{=}C{=}O$$

(xix) Oxymercuration of Olefins

It is a convenient method for the formation of carbon-oxygen bond and is generally performed with a mercury salt e.g. mercuric acetate. It has been shown that any mercury salt can be obtained from mercuric oxide and an organic acid on sonication (J. Einhorn, C. Einhorn and J.L. Luchi, J. Org. chem. in press) Thus, selective reaction of limonene, which is generally difficult, is achieved in excellent yield (80%) under sonication compared to 40% yield under usual conditions.

$$\text{Limonene} \xrightarrow[\text{HgO/t-BuCO}_2\text{H, RT,))))}]{\text{THF — H}_2\text{O(1:1)}} 80\%\ \alpha\text{-Terpineol.}$$

Limonene

80% α-Terpineol.

(xx) Isotopic Labelling

Bromine can be quantitatively exchanged by iodine-123 by sonication of bromo compound (e.g., 17-bromoheptadecanoic acid) with iodine-123 in presence of sodium thiosulphate and butan-2-one at 100°. (J. Mertens, W. Vanrycheghem, A. Bossuyl, P. Vanden Winkel and R. Vanden Drcessche, J. Label Camp. and Radiopharm. 1984, **21**, 843).

$$Br\,(CH_2)_{16}CO_2H \;+\; I^* \xrightarrow[100°,\ ))))\ \text{butan-2-one}]{Na_2S_2O_3} I^*(CH_2)_{16}CO_2H$$

17-Iodo-hepta decanoic acid

(xxi) Grignard Reagents

The treatment on an alkyl or aryl halide in ether with magnesium under sonication in common laboratory cleaning bath gives 90% yield to the grignard reagent. (J.L. Luche and J.C. Damiano, J. Am. Chem. Soc., 1981, **102**, 7925, W. Oppolzer and A. Nakao, Tetrahedron Lett., 1986, **27**, 5471)

$$R - X + Mg \xrightarrow[))))]{\text{ether}} + RMgX \quad 90\%$$

(xxiii) Isomerisation of Maleic Acid to Fumaric Acid

Maleic aicd on sonication in presence of bromine-water gives fumaric acid (I.E. Elpiner, A.V. Sok olskaya and M.A. Margulis, Nature (London), 1965, **208**, 945)

$$Br_2 \xrightarrow{H_2O,\ )))} 2Br^{\bullet}$$

$$\text{HC(COOH)=CH(COOH)} + Br^{\bullet} \longrightarrow \text{Br–CH(COOH)–}\dot{\text{C}}\text{H(COOH)} \xrightarrow{))))} \text{HOOC–CH=CH–COOH} + Br^{\bullet}$$

Maleic acid

Fumaric acid

(xxiii) Dieckmann Cyclisation

The reaction of diethyl adipate with ultrasonically dispersed potassium effects Dieckmann cyclisation to give 2-carbethoxycyclopentanone (J.L. Luche, C. Petrier and C. Duputy, Tetrahedron Left, 1985, **26**, 753)

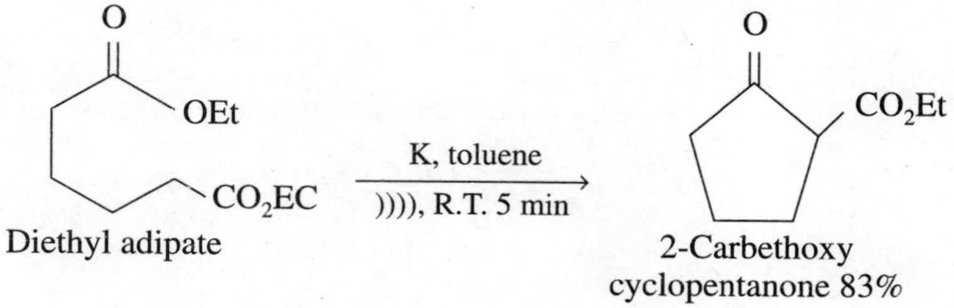

O

OEt

CO_2EC

Diethyl adipate

K, toluene

)))), R.T. 5 min

O

CO_2Et

2-Carbethoxy
cyclopentanone 83%

6.13 CONCLUSION

Ultrasound is a methodoly, which has been used in organic synthesis. It dramatically increases the rates of chemical reaction. It has been possible to effect a large number of organic transformations. The rates of number of organic reactions has increased. Some of the organic reaction which have been Sonochimically induced include Bouveault reaction, cannizzaro reaction, strecker synthesis, Reformatsky reaction, Barbier reaction and Dickmann cyclisation.

Chapter 7

Transformations Using Microwave Irradiations

7.1 ANTHRACENE-MALEIC ANHYDRIDE ADDUCT

The reaction of anthracene with maleic anhydride gives the adduct, **9,10-dihydroanthracene-endo-α, β-succinic anhydride**. This reaction, known as **Diels-Alder reaction** is normally carried out by refluxing the reactants in xylene for about 20 min using an oil bath. It has been found that by using microwave, the reaction can be campleted in about one minute.

Anthracene + Maleic anhydride ⟶ Adduct

Materials

Anthracene	3 g
Maleic anhydride	1.15 g
Diglyme	5 mL

Procedure

Powdered mixture of anthracene (3 g) and maleic anhydride (1.15 g) is mixed with diglyme (5 mL). The mixture is heated in a beaker (covered with a watch glass) in a microwave over for about 60 sec. at a medium power level. The reaction mixture is cooled to room temperature and the separated adduct is filtered and washed with ethanol, yield 3.3 g. M.P. 262-63° (decomp.).

Notes

1. Diels-Alder reaction can also be conducted in water (R. B. Woodward and H. Baer, J. Am. Chem. Soc., 1948, **70**, 1160; R. Breslow and D. Rideout, J. Am. Chem. Soc, 1980, **102**, 7816). See the preparation of endo-cis-1,4-endoxo-Δ^5-Cyclohexene 2,3-dicarboxylic acid (Section 2.17)
2. The Diels-Alder reaction of anthracene with dimethyl fumarate gives the adduct in good yield (R.J. Giguere, T.I. Bray, S.M. Durean and G. Majeticn, Tetrahedron lett., 1986, **27**, 4945)

Anthracene + Dimethyl fumarate (MeO_2C–CH=CH–CO_2Me) $\xrightarrow[\text{p-xylene, 10 mm}]{\text{MW}}$ Adduct 87% (CH—CO_2Me / CH—CO_2Me)

7.2 ASPIRIN (ACETYLSALICYLIC ACID)

It is an analgesic drug and is obtained by the acety lation the salicylic acid with acetic anhydride in presence of catalytic amount of conc. sulphuric acid.

Salicylic acid (OH, C(=O)—OH) + $(CH_3CO)_2O$ (Acetic anhydride) $\xrightarrow{H^+}$ Aspirin (Acetyl Salicylic acid) (O—C(=O)—CH_3, C(=O)—OH) + CH_3COOH (Acetic acid)

Mechanism

Acetic anhydride (CH_3—C(=O)—O—C(=O)—CH_3) $\xrightarrow{H^+}$ CH_3—C(=$\overset{\oplus}{O}$—H)—O—C(=O)—CH_3; R—$\ddot{O}$—H $\rightleftharpoons$ CH_3—C($\ddot{O}$—H)($\overset{+}{O}$(R)H)—O—C(=O)—CH_3 ↓ $-CH_3$—C(=O)—OH

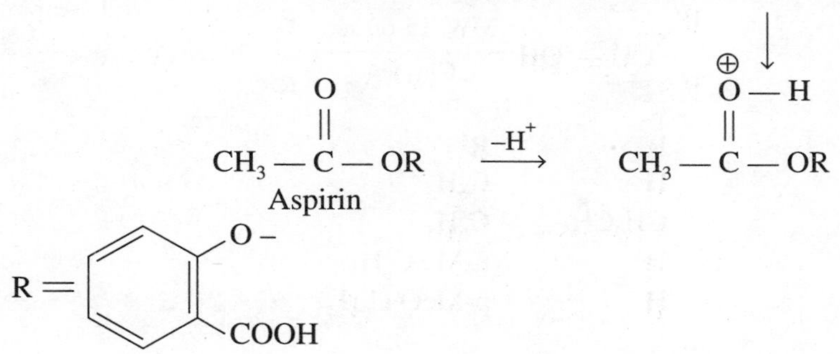

$$CH_3-\overset{\overset{O}{\|}}{C}-OR \xrightarrow{-H^+} CH_3-\overset{\overset{\overset{\oplus}{O}-H}{\|}}{C}-OR$$

Aspirin

R = 2-(O–)C₆H₄COOH (o-carboxyphenyl, –O– attached to ring bearing COOH)

Materials

Salicylic acid	2.5 g
Acetic anhydride	5 mL
Conc. Sulphuric acid	1-2 drops

Procedure

A mixture of dry salicylic acid (2.5 g, 0.018 mol), distilled acetic anhydride (5 mL) and conc. H_2SO_4 (1-2 drop) is heated (in a porcelin dish) in a miorowave oven for 15-20 seconds. The mixture is cooled, water (50 mL) added and stirred. The separated aspirin is filtered and crytallised from dilute alcohol, m.p. 135-136° yield 2.9 g (90%)

1. Using the same procedure, hydroquinone diacetate, m.p. 122° (94.8% yield) is prepared from hydroquinone, actic anhydride and 1-2 drops of conc. H_2SO_4
2. Benzoquinone can also be converted into hydroquinone diacetate by reductive acetylation. The method involves heating a mixture of p-benzoquinonone, acetic anhydride, anhydrous Sodium acctate and Zn dust (30 sec heating in a MW oven) and working up in the usual way. By this procedure hydroquinone deacetate m.p. 122° is obtained in 67% yield.

7.3 BENZALDEHYDE

Normally aldehydes are obtained by the oxidation of primary alcohols by various oxidising agents, like potassium permanganate, chromium trioxide, potassium chromate and potassium dichromate. However the best procedure is by using clayfen in solid state, a rapidly and high yield of the synthesis of carbonyl compound is achieved. Thus, benzaldehyde is obtained by heating benzyl alcohol with clayfen (a clay material, commercially available)

$$C_6H_5CH_2OH \xrightarrow[\text{15-60 Sec}]{\text{Clayfen, MW}} C_6H_5CHO$$

Using this procedure a number of aldehydes and ketones using appropriate starting 1° or 2° alcohols can be obtained.

$$\begin{matrix} R' \\ \quad \diagdown \\ \quad CH-OH \\ \quad \diagup \\ R \end{matrix} \xrightarrow[\text{Clayfen}]{\text{MW, 15-60 sec}} \begin{matrix} R \\ \quad \diagdown \\ \quad C=O \\ \quad \diagup \\ R' \end{matrix}$$

80-97%

R	R′
H	C_6H_5
CH_2CH_3	C_6H_5
H	p-Me C_6H_4
H	p-MeO C_6H_4

Notes

1. R.S. Varma and K. Dahiya, Tetrahedron Lett., 1997, **38**, 2043
2. In the oxodation of 1° alcohols there is exclusive formation of aldehydes with no formation of the corresponding carboxylic acids.
3. Primary and secondary alcohols can also be oxidised to the corresponding carbonyl compounds by using 35% MnO_2 doped silica under M.W irradiation conditions (R.S. Varma, R.K. Saini and R. Dahiya, Tetrahedron Lett. 1997, **38**, 7823

$$\begin{matrix} R_1 \\ \quad \diagdown \\ \quad CHOH \\ \quad \diagup \\ R_2 \end{matrix} \xrightarrow[\text{MW}]{MnV_2-\text{silica}} \begin{matrix} R_1 \\ \quad \diagdown \\ \quad C=O \\ \quad \diagup \\ R_2 \end{matrix}$$

4. Chromium trioxide (CrO_3) impregnated on pre-moistened alumina is another oxidising agent which oxidises benzyl alcohols by simple mixing at room temperature. The reactions are relatively clean with no tar formation. Also, this in no overoxidation to carboxylic acids (R.S. Varma and and R.K. Saini, Tetrahedron Lett., 1998, **39**, 1481). The mole ratio of substrate and reagent used is 1:2, except is case to benzoin where a mole ratio of 1:3 is used

$$\begin{matrix} R_1 \\ \quad \diagdown \\ \quad CH-OH \\ \quad \diagup \\ R_2 \end{matrix} \xrightarrow[\text{MW}]{\text{Wet } CrO_5-Al_2O_3} \begin{matrix} R_1 \\ \quad \diagdown \\ \quad C=O \\ \quad \diagup \\ R_2 \end{matrix}$$

R_1 = Ph; R_2 = H, Ph, Me, PhCO
R_1 = 4-Me C_6H_4, 4-MeO C_6H_4, 4-NO_2 C_6H_4; R_2 = H

5. Primary and secondary alcohols can also be oxidised to the corresponding carbonyl compounds by oxidation with iodobenzene diacetate (IBD) supported on alumina and subjecting to irradiation with microwaves (R.S. Varma, R. Dahiya and R.K. Saini, Tetrahedron Lett., 1998, **38**, 7029.

$$R_1R_2CH-OH \xrightarrow[\text{MW}]{\text{1BD — Alumina}} R_1R_2C=O$$

R_1 = Ph; R_2 = H, Et, PhCO

R_1 = 4-Me C_6H_4, 4-OMe C_6H_4; R_2 = H

7.4 BENZIMIDAZOLE

It is prepared by heating a mixture of o-phenylenediamine with formic acid.

$$\text{o-Phenylene diamine} + \underset{\text{Formic acid}}{HCO_2H} \xrightarrow[\text{MW, 3 min}]{\Delta} \text{Benimidazole}$$

Materials

o-Phenylene diamine	2.7 g
Formic acid (90%)	1.8 g (1.6 mL)

Procedure

A mixture of o-phenylene diamine (2.7 g, 0.025 mol) and formic acid (90%, 1.8 g, 1.6 mol) is heated in a microwave oven for 3 min. The resulting solution to cooled and mixture rendered alkaline by addition of sodium hydroxide solution. The separated benzimidazole is washed with cold water and crystallised from hot water. The yield is 2.2 g (56%), m.p. 176°.

Notes

1. o-Phenylene diamine is obtained by reduction of o-nitro aniline with zinc and sodium hydroxide as follows: A mixture of o-nitro aniline (4.6 g, 0.033 mol) and sodium hydroxide solution (20%, 2.7 mL) and ractified spirit (15 mL) is stirrred and refluxed (water-bath). The source of heat is removed and zinc powder (9 g, 0.14 mol) is added in small lots (2-3 g. each time). The mixture is refluxed for 40 min and filtered. To the filtrate is added sodium dithionite (0.2 g), solution concentrated under reduced pressure (steam bath) to about 8-10 mL. On cooling (ice-salt mixture), o-phenylene diamine separates out. It is crystallised from hot water. Yield 2.84 g (79%), m.p. 100-101°.
2. Using acetic acid in place of formic acid gives 2-methylbenzimidazole in 56% yield, m.p. 176°. And using phenylacetic acid in place of formic acid gives 2-benzylbenzimidazole in 55% yield, m.p. 191°.
3. Benzimidazoles can also be prepared by condensation reaction of ortho-esters with o-phenylenediamine in the presence of KSF clay either under refluxing conditions using

focused microwave irradiation (D. Villemin, M.Hammadi and B. Marlin, synth. Commun., 1996, 26, 2895.

$$(EtO)_3C{-}H + C_6H_4(NH_2)_2 \xrightarrow[\text{microwave}]{\text{KSF clay}} \text{Benimidazole}$$

Benimidazole

7.5 BENZOPHENONE

It is obtained by the oxidation diphenyl methyl bromide with sodium hypochlorite using ultrasound.

$$Ph_2CHBr \xrightarrow[\text{2 hr))))}]{NaOCl,\ CH_3CN} Ph_2C=O \quad 93\%$$

Procedure

A solution of diphenyl methyl bromide (a benzylic halide) (1 m mol) in acetonitrile (15 mL) and sodium hypochlorite (20 m mol of 0.05 N NaOCl solution) was exposed to ultrasound (sonicator bath). The progress of the reaction was monitered by TLC. After the starting bromide disappears (about 2 hr), neutralise the reaction mixture with 1N-H_2SO_4. Extract with dichloromethane (3 × 10 mL), dry the extract ($MgSO_4$) and distil to obtain benzophenone, m.p. 48-49°, yield 93%

Notes

1. J.M. Khurana, P.K. Sahoo, S.S. Titus and G.C. Maikap, Synthetic Communication, 1990, **20**, 1357.
2. Soclium hypochlorite oxidises secondary benzylic halides to ketones. However, primary benzylic halides are oxidised to aldehydes and carboxylic acids.
3. It is believed that the oxidation takes place in two steps. The first is the nucleophilic substitution of the halide in a slow step to give the corresponding alcohol, which undergoes subsequent faster oxidation to the product.

$$RR'CHX \xrightarrow{NaOCl,\ CH_3CN} RR'CHOH$$

$$RR'CHOH \xrightarrow{NaOCl,\ CH_3CN} RR'CO + R\,COOH \quad (\text{if } R' = H)$$

4. Benzophenone can also be prepared in 98% yield by the oxidation of α-phenyl benzyl amine with NaOCl in presence $Bu_4 N^+ X^-$ (G.A. Lee and H.H. Frecdman, Tetrahedron Lett., 1976, 1641)

$$C_6H_5-\underset{\underset{C_6H_5}{|}}{CH}-NH_2 + NaOCl \xrightarrow[\text{EtOAc 140 min}]{Bu_4\overset{+}{N}\overset{-}{X}} C_6H_5COC_6H_5$$

α-Phenyl benzyl amine → 98% Benzophenone

7.6 BENZYLIDENEANILINE

It is prepared by the condensation of benzaldehyde with aniline. The reaction is conducted in presence of catalytic amount of montmorillonie K10 clay in microwave over for 1 min.

$$C_6H_5CHO + H_2NC_6H_5 \xrightarrow[\text{1 min}]{\text{K-10 clay, MV}} C_6H_5CH=NC_6H_5$$

Benzaldehyde + Aniline → Benzylidene aniline

Procedure[1]

A mixture of equimolar amounts of benzaldehyde and aniline is heated in a microwave oven for 1 min in presence of catalytic amount of montmorillonite K10 clay. The formed benzylideneaniline is crystallised from carbon disulphide and melted at 48°; the m.p. increased to 56°, if the melted product is solidified and the mp determined, yield 97%

1. R.S. Varma and R. Dahiya, Synlett, 1997, 1245.
2. Normally the imines are prepared by the condensation of carbonyl compounds and amines followed by the azeotropic removal of water form the intermediate that is the driving force for the reaction. The reaction is normally catalysed by p-toluene sulphonic acid, molecular sieve and titanium chloride using Dean Stark's apparatus which requires excess of solvents like benzene or toluene. The above procedure is a very converent procedure.
3. Using the above procedure, a number of substituted benzaldehydes (like 2-OH, 4-OH, 4-Me, 4-NMe_2 and 4-Me could be converted into the corresponding imines (1-3 min heating in MW oven) in 90-96% yields

7.7 2, 3-DIPHENYLQUINOXALINE

It is prepared by heating o-phenylenediamine and benzil in alcohol.

$$\text{o-Phenylenediamine} + \text{Benzil } (C_6H_5-CO-CO-C_6H_5) \xrightarrow[\text{1 min}]{\text{MV}} \text{2,3-Diphenyl quinoxaline}$$

Materials

o-Phenylene diamine	1.65 g
Benzil	3.15 g
Ethanol	15 mL

Procedure

A solution of o-phenylenediamine (1.65 g, 0.015 mol) and benzil (3.15 g, 0.015 mol) in ethanol (15 mL) is heated in a micowave oven for 1 minute. The solution is cooled and the separated product crystallised from alcohol, m.p. 125°-126°, yield 2.75 g (51%).

Notes

1. o-Phenylenediamine is obtained as given in the preparation of benzimidazole (Section 7.4).
2. Using 9, 10-phenanthracene in place of benzil in the above preparation yields 1,2: 3,4-dibenzophenanazine in about 50% yield, m.p. 125-126°
3. Using phenylpyruvic acid in place of benzil in the above preparation gives 3-benzyl-2-oxo-1,2-dihydroquinoxaline in 54% yield, m.p-203°.

7.8 ETHYL BENZOATE

It is obtained by the **esterification** of benzoic acid with ethyl alcohol in presence of catalytic amount of conc. sulfuric acid using sonication.

$$\underset{\text{Benzoic acid}}{C_6H_5COOH} + C_2H_5OH \xrightarrow[\text{)))) 3 hr}]{\text{Conc } H_2SO_4} \underset{\text{Ethyl benzoate 89\%}}{C_6H_5COOC_2H_5} + H_2O$$

Procedure[1]

Dissolve benzoic acid (10 m mol) in dry ethanol (15 mL). Add sulphuric acid (0.5 mL, 10 N). Expose the reaction mixture to ultrasound. Monitor the completion of the reaction by TLC. Work up the reaction by addition of water (25 mL) and extraction with dichloromethane (2 × 25 mL).

Wash the organic layer with sodium bicarbonate (10%, 2 × 15 mL) and water (2 × 15 mL). Dry the dichloromethane extract over anhydrous magnesium sulphate and evaporate the solvent. Ethyl benzoate is obtained in 89% yield, b.p. 211-213°.

Notes

1. J.M. Khurana, P.K. Sahoo and Y.C. Maikap, Synthetic Communication, 1990, **20**, 2267
2. Using the above procedure, esterification of aromatic, aliphatic, α, β-unsaturated, mono and dicarboxylic acids has been successfully archieved. The methyl-, ethyl-, n-propyl-, isopropyl and n-butyl alcohols, gave high yields of esters. Even t-butyl esters have been obtained in reasonable yields.

7.9 INDOLE

It is obtained by intramolecular cyclisation of n-formyl-o-toluidine with a strong base like sodium methoxide at high temperature. This procedure is known as **Madelung Indole Synthesis** (M. Madelung Ber., 1912, **45**, 1128)

CH_3 / N·CHO / H
N-Formyl o-toluidine
NaOMe, Δ →
$CH_2^{\ominus}$ / O / C — H / N / H
H^+, H_2O →
N / H
Indole

Materials

Sodium	0.95 g
Methanol (anhyd.)	25 mL
Pot. acetate (fused)	8 g
N-Formyl-o-toluidine	5.66 g

Procedure

N-formyl-o-toluidine (5.66 g, 0.42 mol) and freshly fused and powdered potassium acetate (8 g, 0.85 mol) is added to a solution of sodium methoxode (prepared by adding sodium (0.95 g, 0.042 mol) to anhydrous methanol (25 mL). The mixture is refluxed to get a clear solution. Excess methanol is removed (in vacuo) and the reaction mixture heated at 300-350° using wood's metal bath. The heating is continued for about 30 min. Alternatively, the mixture is heated in a microwave oven for 5 min. Excess o-toluidene is removed (by appling vaccum), reaction mixture cooled and water (15 mL) added. Excess o-toludine is removed (steam distillation). Cooling of the reaction mixture gives indole, which is filtered and crystallised from petroleum ether, yield 0.8 g (17 %), m.p. 48-50°.

Notes

1. The required N-formyl-o-toluidine is obtained by refluxing a mixture of o-toluidine (10.3 mL, 0.1 mol) and formic acid (90%, 5.25 g, 0.1 mol). The reaction mixture is distilled to give N-formyl-o-toluidine, b.p. 173.75/25 mm). It solidifies on colling, yield of 13 g (80 %) m.p. 57-59°.
2. Fused potassium acetate is obtained by heating it in a porcelain dish until it melts. It is kept in molten state for 2-3 min., cooled (desiccator) and powdered.
3. See also preparation of indole by the decarboxylation of indole 2-carboxylic and by heating in MW over at 55° (Section 2.25)

7.10 4-METHYLCARBOSTYRIL (2-HYDROXY 4-METHYL QUINOLINE

It is obtained by the cyclisation of acetoacetanilide with conc. H_2SO_4

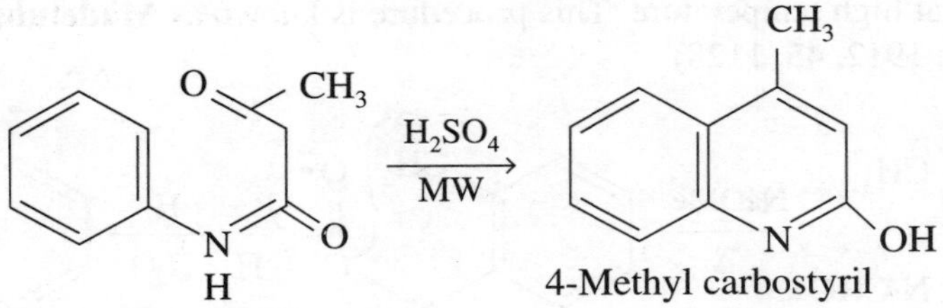

H_2SO_4 / MW

Acetoacetanilide → 4-Methyl carbostyril

Procedure

A mixture of acctoaetanilide (8.9 g, 0.05 mol) and conc. H_2SO_4 (10 mL, added slowly) is heated in a microwave oven for 60 sec. The reaction mixture is cooled and water (250 mL) added with stirring. The separated 4-methyl carbostyril is filtered, washed with water and crystallised from methanol, m.p. 223-224°, yield 7 g (87%)

7.11 4-METHYL-7-HYDROXYCOUMARIN

It is obtained by the **Pechmann reaction** of resorcinol with ethylacetoacetate (H.V. Pechmann and C. Duisberg, Ber., 1883, 16, 2119)

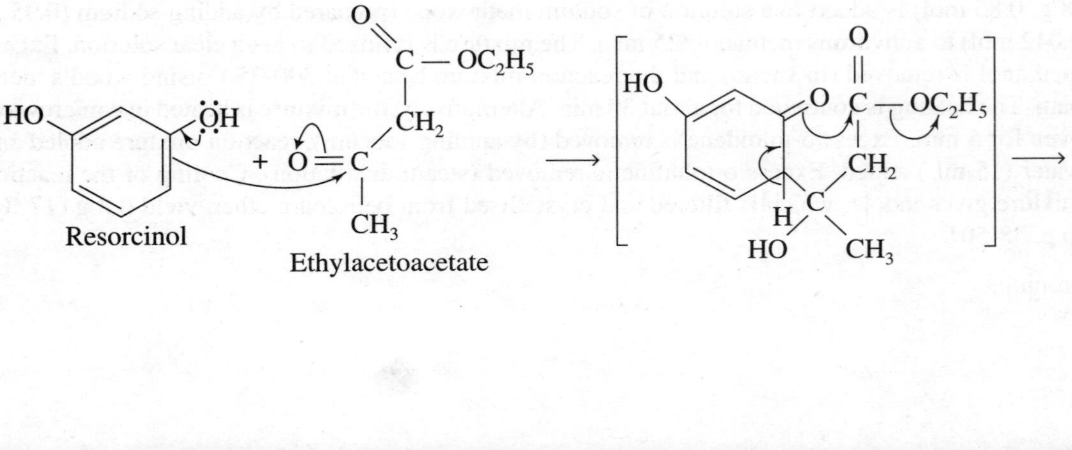

Resorcinol + Ethylacetoacetate →

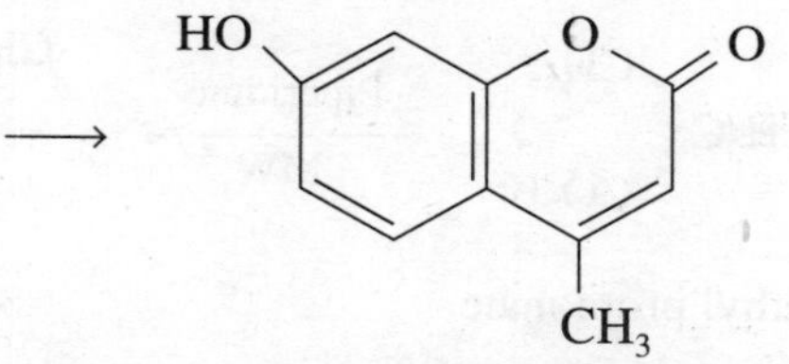

→ HO, O, O, CH_3

4-Methyl-7-hydroxycoumarin

It has been found that the reaction proceeds very will by microwave irradiation (V. Singh, J. Singh, P. Kaur and C.L. Kad, J. Chem. Res. (S), 1997, 58)

Materials

Resorcinol	11 g
Ethylacetoacetate	13 g
Conc. H_2SO_4	13 g

Procedure

To a solution of resorcinol (11 g, 0.1 mol) in ethyl acetoacetate (13 g, 0.1 mol) is added cautiously conc. H_2SO_4 (13 g). The reaction mixture is heated in a MW oven for 2 min. It is poured into ice, separated produced filtered and crystallised from dilute alcohol, m.p. 185-186°, yield 16 g (91%).

Notes

1. Using the above procedure a number of substituted 4-methyl coumarins could be prepared.
2. Pechmann condensation can also be performed by heating in ionic liquid at 80° for 15 min (G. Gu, F. Shi and Y.J. Deng, J. Mol. Catal. A. Chem, 2004, **212**, 71)

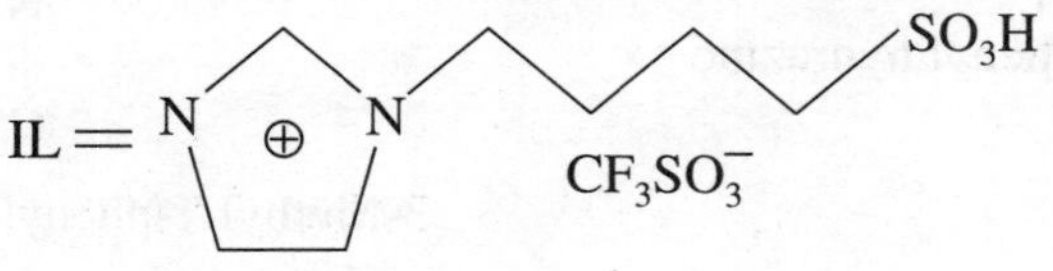

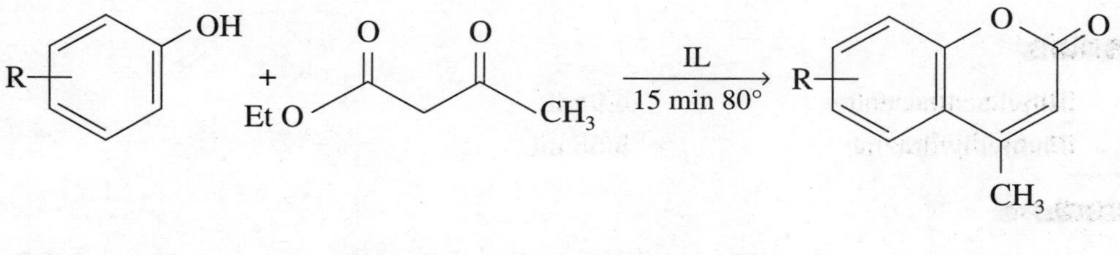

IL = N ⊕ N, SO_3H, $CF_3SO_3^-$

R, OH + O, O, Et O, CH_3 $\xrightarrow[\text{15 min 80°}]{\text{IL}}$ R, O, O, CH_3

7.12 3- METHYL-7-METHOXY COUMARIN

It is obtained by **Knoevenagel condensation** of 2-hydroxy-4-methoxy benzaldehyde with ethyl propionate in presence of piperidine by microwave irradiation.

2-Hydroxy-4-methoxy benzaldehyde + Ethyl propionate $\xrightarrow[\text{MW}]{\text{Piperidine}}$ 3-Methyl 7-methoxycoumarin

A mixture of 2-hydroxy-4-methoxy benzaldehyde and ethyl propionate in the molar ratio 1 : 1.2 is heated in presence of piperidine in a microwave oven for 2 min. The product obtained is crystallised from dilute alcohol. Determine its m.p. and yield. Record its NMR spectra.

Note

1. This is a general procedure for the synthesis of substituted 3-methyl coumarins.
2. The procedure is a modification of the procedure of D. Bogdal, J. Chem. Res. (S), 1998, 468.

7.13 3-METHYL-1-PHENYL-5-PYRAZOLONE

It was earlier prepared by the condensation of ethyl acetoacetate with phenyl hydrazine by heating at 115-120° in an oil bath for 3 hrs. It is found (D. Villenmm and B. Labiad, Synthetic Commun., 1990, **20**, 3213) that the reaction takes a much shorter time (about 10 min) by heating in a MW oven.

$CH_5-CO-CH_2-CO-OEt$ Ethylacetoacetate + $C_6H_5NHNH_2$ Phenyl hydrazine $\xrightarrow[\text{10 min}]{\text{MW}}$ 3-Methyl-1-phenyl pyrazolone 90%

Materials

Ethyl acetoacetate	4.9 mL
Phenyl hydrazine	3.65 mL.

Procedure

A mixture of ethylacetoacetate (4.9 mL, 0.0384 mol) and phenyl hydrazine (3.65 mL, 0.037 mol) is heated in a conical flask is a microwave oven (280 W) for 10 min. The reaction mixture is cooled, separated product filtered and crystallised from dilute alcohol (1:1) to yield 3-methyl-1-phenyl-5-pyrazolone, m.p. 129°, yield 6.5 g (100%).

7.14 2-PHENYLINDOLE

It is obtained by heating acetophenone phenyl hydrazone with polyphosphoric acid. This procedure is known as **Fischer Indole Synthesis**. (E. Fisher and F. Jourdan, Ber., 1883, **16**, 2241)

H_3C C_6H_5 C N N H — Acetophenone phenyl hydrazone ⇌ H_2C C_6H_5 C N—H N H $\xrightarrow{H^+}$ H_2C C_6H_5 C NH_2 N ⊕ H ⟶

H CH_2 C_6H_5 C NH_2 N ⊕ H ⟶ CH_2 C_6H_5 C NH_2 NH_2 + ⟶ H C_6H_5 N NH_2 H ⟶

$\xrightarrow{-NH_3}$ N H C_6H_5

2-Phenyl indole

Materials

Acetophenone phenylhydralzone	7 g
Polyphosphoric acid	40 g

Procedure

A mixture of acetophenone phenylhydrazone (7 g) and polyphosphoric acid (40 g) is heated in a microwave over for 3 min. Water (100 ml) is added to the reaction mixture, separated product filtered and washed with water. It is crystallised from alcohol, m.p. 187°, yield 5 g (79%)

Notes

1. Acetophenone phenyl hydrazone is prepared by heating a mixture of acetophenone (10 g, 0.835 mol), phenyl hydrazine (9 g, 0.835 mol) and alcohol (30 mL) containing 2-3 drops of glacial acetic acid in a microwave oven for 2 min. The separated phenyl hydrazone is

washed with dil. HCl, ethanol (2-5 mol) and crytallised form ethanol, m.p. 106°, yield 11 g.

2. Polyphosphoric acid is obtained by dissolving phosphorus pentoxide (25 g) in commercial orthophosphoric acid (d. 1.75, 15 g).
3. Using phenyl hydrazone of acetone, in place of acetophenone phenyl hydrazone, 2-methylindole can be obtained by heating with anhyd. $ZnCl_2$ in a microwave oven for 3 minutes. The products, m.p. 95° is obtained in 50% yield.
4. 2,3-Dimethyl indole can be obtained by heating phenyl hydrazine and ethyl methyl ketone in water at 270° in a microwave over for 30 min in 64% yield (J.M. Kremsner and C.O. Kappe, Eur, J. Org. Chem., 2005, 3672)

Phenyl hydrazine + Ethyl methyl ketone $\xrightarrow[\text{MW, 270°, 30 min}]{H_2O}$ 2,3- Dimethyl indole 64 %

Alternatively, the reaction can be conducted in a sealed tube by heating at 290° for 30 min.

7.15 PIPERAZINE-2,5-DIONE (DIKETOPIPERAZINE)

It is obtained by heating glycine in ethylene glycol as solvent.

$$2\ NH_2CH_2COOH \xrightarrow[\text{MW, 5 min}]{\text{Ethylene glycol}} \text{Diketopiperazine}$$

Glycine

Materials

Glycine	5 g
Ethylene glycol	25 mL

Procedure

A miture of glycine 5 g (0.09 mol) and ethylene glues (25 mL) is heated in a microwave oven for 5 min. The resultant mixture is coold and left in a refrigerator for 24 hr. The separated product is separated by decantation and crystallised from hot water, m.p. 310-12° (decompn.) yield 1.7 g (45%)

Notes

1. Boil a few mg. of the product (obtained above) with ninhydrin in 0.5 mL water. Absence of blue-purple colour indicates that the product is free from the byproduct (peptide material).

7.16 PHENYLACETIC ACID

It is obtained by hydrolysis of benzyl cyanide with conc. H_2SO_4.

$$\underset{\text{Benzyl cyanide}}{C_6H_5CH_2CN} + H_2SO_4 \xrightarrow{H_2O} \underset{\text{Phenylacetic acid}}{C_6H_5CH_2COOH} + (NH_4)_2SO_4$$

Materials

Benzyl cyanide	5.4 mL
Conc. H_2SO_4	5 mL
Glacial acetic acid	5 mL

Procedure

A mixture of benzyl cyanide (5.4 ml, 0.047 mol), water (10 mL), glacial acetic acid (5 mL) and conc. H_2SO_4 (5 mL) is heated in a beaker (covered with a watch glass) for 1 min in a MW oven. The mixture is poured into cold water (25 mL). The separated phenylacetic acid is filtred and crystallised form hot water, yield 2.4 g (37.7%) m.p. 77°.

Notes

1. If impure, phenyl acetic acid is purified by dissolving in sodium carbonate solution and acidifying the clear filtrate.
2. Using the same procedure, **p-toluic acid**, m.p. 178° can be obtained in about 50% yield from p-tolunitrile.

7.17 PHENYL PROPYL THIOETHER

It is prepared by the reaction of thiophenol with n-propylbromide in acetone solution using sonication.

$$\underset{\text{Thiophenol}}{C_6H_5SH} + \underset{\text{n-Propyl bromide}}{BrCH_2CH_2CH_3} \xrightarrow[)))))]{\text{Acetone}} \underset{\text{Phenyl propyl thoether (89\%)}}{C_6H_5SCH_2CH_2CH_3}$$

Procedure[1]

Thiophenol (0.005 mol), n-propyl bromide (0.005 mol) and anhydrous potassium carbonate (0.0075 mol) in dry acetone (20 mL) or dimethylformamide on sonication for one hr gives 89% yield of phenylpropyl sulphide. In place of sonication, even stirring with a magnetic stirrer is good. However, sonication acclerates the reaction. The reaction is continued till the starting thiophenol disappeared as seen by TLC. After the completion of the reaction, the mixture was filtered, the combined filtrate concentrated and ice-cold water (50 mL) added. Extract the product with dichloromethane, dry (anhyd. $MgSO_4$) and concentrate to give the required phenyl propyl thioether in 89% yield.

Notes

1. J.M Khurana and P.K. Sahoo, Synthetoc Communication 1992, 22, 1691
2. The reaction can be conducted eithr by stirring or by sonication; the later techniques acclerates the reaction.
3. Using the above procedure a number of thiophenols (like thiophenol, 2-mircapto benzothiazol, benzyl mercaptan, 2-mercaptoethanol, 1-butanethiol) could be reacted with various halides (like propyl bromide, n-hexadecyl bromide, n-butyl chloride, benzyl chloride) to give the corresponding thioethers.
4. In certain cases, small amount of disulphides are also obtained (as seen by TLC)
5. The yields of thioethers may be improved by phase transfer catalysts [A.W. Herriott and D. Picker, Tetrahedron Lett., 1972, 4521; Synthesis, 1975, 447; J. Am. Chem Soc, 1975, **97**, 2345; K.C. Majundar, A.T. Khan and S. Saha, Indian J. Chem Sect. B, 1991, **30**, 643].

7.18 1,2,3,4-TETRAHYDROCARBAZOLE

It is prepared by heating a mixture of cyclohexanone, glacial acetic and phenyl hydrazine.

$NHNH_2$ + O (HOAc, MW 3 min) →

Phenyl hydrazine

Cyclohexanone

Phenylhydrazone of cyclohexanone

→

1,2,3,4-Tetrahydrocarbazole

Materials

Cyclohexanone	2.45 g
Actic acid (glacial)	9 mL
Phenyl hydrazine	2.45 mL

Procedure

A mixture of cyclohexanone (2.45 g, 0.025 mol), glacial acetic acid (9 mL) and phenyl hydrazine, (2.45 mL, 0.025 mol) is heated in a microwave over for 3 min. The solution on cooling (ice-salt mixture) gives 1,2,3,4-tetrahydrocarbazole, which is filteral, washed with water and crystallised from methanol, m.p. 116-117°, yield 3.25 g (76%)

7.19 MISCELLANEOUS APPLICATIONS OF MICROWAVE REACTIONS

(i) Deactylation of Benzaldehyde Diacetates

The aldehydic groups are generally protected by acetylation (to give the corresponding diacetates). The conventional deportection of acetyls is carried out either by using sodium hydroxde or potassium carbonates in aqueous tetrahydrofuran solutions left overnight to give the product in moderate yield. A convenient procedure is to carry out the deacetylation using microwaves. In this procedure, the diacetates are absorbed on neutral alumina surface (The diacetate in dichloro methane is mixed with neutral alumina and the solvent evaporated). The absorbed material is heated in a microwave oven by placing in a beaker which in turn is placed in an alumina bath in the microwave oven. The deacetylation is complete in 40 seconds. The product is isolated by extraction with dichloro methane and evaporation of the solvent. The yield is quantitative (R.S. Varma, A.K. Chatteriee and M. Varma, Tet. Lett, 1993, **34**, 3207)

(ii) Deacetylation of Acetyl Derivatives of Phenols and Alcohols

Normally hydroxy groups are protected by acetylation and deacetylation effected by heating with alkali. Using microwave oven, the acetates can be deacetylated by absorbing on to neutral alumina and then heating the absorbed material is a microwave oven (R.S. Varma, M. Varma and A.K. Chatterjee, J. Chem. Soc, Perkin Trans-1, 1993, 999).

(iii) Deportation of Benzyl Esters

The carboxylic functionality is generally protected by converting it into benzyl esters. Standard methods of debenzylation include catalytic hydrogenation and other methods employing, K_2CO_3, $AlCl_3$, Na-NH_3 etc. However, these procedures give low yields and poor chemo selectivity and longer reaction time is required. Using microwaves irradiation, deportation at the benzyl ester can be conveniently effected under mild and solvent free conditions on alumina surface. In this procedure (R.S Varma, A.K Chatterjee, M.Varma, Tetrahedron Lett., 1993, **34**, 4603), the benzyl ester is dissolved in dichloromethane (minimum volume) and mixed thoroughly with acidic alumina. The absorbed material is dried in a beaker in air and irradiated with microwave (for this purpose the beaker containing the adsorbed material is placed in an alumina bath (heat sink) inside the MW oven (130-140°)). The debenzylation is complete in 3-10 min. After competition of the reaction (as indicated by TLC) the product is extracted with hexane-ether (4:1) and than with methanol-dichloromethane (4:1). Methanol is removed under vacuo to give about 95% yield to the carboxylic acid. The hexane-ether extract removes only the byproduct (benzyl alohol)

(iv) Deportation of t-butyldimethylsily Ethers

The hydroxy functionality is generally protectect by tert-butylsilyl group (TBDMS). This group can easily deprotected on alumina surface without using any solvent by microwave irradiation at 75-85° in a microwave oven for about 10 min (R.S. Varma, J.B. Lamture and M. Varma, Tetrahedran Lett, 1993, **34**, 3029) in 80-90 percent yields. The procedure involves mixing (by stirring) the t-butyldimethylsilyl ether dissolved in minimum amount of dichloromethane and neutral alumina. The absorbed material is dried in air (beaker) and placed inside an alumina bath (heat sink) inside the MW oven. After 10 min of irradiation the product is extracted with dichloromethane, solved removed and the residue crystallized from methanol-dichloromethane to give pure hydroxy compound in 80% yield.

(v) Decarboxylation of Aromatic Carboxylic Acids

Aromatic carboxylic and can be conveniently decarboxyled by microwave irradiation using quinoline as solvent in about 90% yield (about 10 min. irradiation in MW oven). The decarboxylated product is isolated by dissolving in ethyl acetate, washing the extract with dilute HCl (1%), water and sodium hydroxide and distillation of the solvent (G.S. Jones and B.J. Chapman, J.Org. Chem, 1993, **58**, 5558)

(vi) Sponification of Esters

Sponification of esters in normally carried by heating with alkali for about 2-3 hrs. However, sponification of even hindered esters can be achieved in few minutes under MW irradiation using solid-liquid PTC conditions without solvent. (A. Loupy, P. Pigeon, M. Ramdani and P. Jacqualt, Synth, Commun. 1994, **24**, 159

$$R-\overset{\overset{\large O}{\|}}{C}-OR' \xrightarrow[\text{(2) } H^+]{\text{(1) KOH-Aliquat, 3-4 min MW}} R-\overset{\overset{\large O}{\|}}{C}-OH$$

(vii) Dethioacetalization of Thioacetals and Thioketals of Aldehydes and Ketones

Thioacetals and thioketals of aldehydes and ketones are usually deprotected by using toxic heavy metal catalysts and as Hg^{2+}, Ag^{2+}, Ti^{4+}, Cd^{2+}, Te^{3+} or reagents like benzenselenic anhydrides. A convenient way for dethioacetalization has been achieved in solid state using clayfen and subjecting to MW irradiation (40 seconds) (R.S. Varma, and A.K. Chatterjee, J.Chem. Soc., Perkin Tran.I, 1993, 999)

$$R_1R_2C(S-R_3)(S-R_4) \xrightarrow[\text{MW, 40 sec}]{\text{Clayfen}} R_1R_2C=O$$

about 90%

(viii) Deprotection of Oximes

Oximes are used as protecting groups for carbonyl compounds. The deoximation was earlier effected by reagents like Raney nickel, pyridinum chlorochromate, dinitrogen tetroxide etc. It has been possible to effect solvent-free deportation of oximes using benin ammonium presulphate on silica and subjecting to MW irradiation for 1-2 min (R.S Varma and H.M. Meshram, Tetrahedron Lett. 1997, **38**, 5427).

$$R_1R_2C=N-OH \xrightarrow[\text{MW, 1-2 min}]{(NH_4)_2S_2O_8-\text{Silica}} R_1R_2C=O$$

70-80%

Oximes can also be deprotected by ultra sonically stimulated Bakers yeast (A. Kamal, M.W. Rao and H.M. Meshram, J. Chem. Soc. Perkin Trans I, 1991, 2056). The procedure involves incubation of a mixture of ultrasonically pretreated bakers yeast (15 g) in phosphate buffer (pH 7.2; 150 ml) and the oxime (of p-methoxybenzaldehyde) at 37° per 2-3 days. The mixture was filtered and extracted with ethylacetale. The organic phase is dried and evaporated (in vacco) to give the a aldehyde (p-methoxybenzaldehyde, b.p 247°, yield 98%), which was purified by column chromatography.

(ix) Cleavage of Semicarbazones and Phenylhydrazones

Semicarbazones and phenylhydrazones can be conveniently eleaved to give the original carbonyl compounds by using ammonium pesulphate impregnated on montmorillonite K 10 clay and subjecting to MW irradiation (R.S. Varma and H.K. Meshram, Tetrahedron Lett, 1997, **38**, 7913; R.S. Varma and D. Kumar, Synth. Commun, 1999, **29**, 1333).

$$R_1R_2C=N-NH-R \xrightarrow[(NH_4)_2S_2O_8-\text{Clay}]{\text{MW}} R_1R_2C=O$$

(x) Wittig Reaction

In the microwave assisted Wittig reaction triphenylphosphine is reacted with organic halides to give the phosponium salt, which on treatment with base give the ylide; finally the ylide on reaction with ketone yielded the olefins (A. spinella, T. Fortunati and R.Sorienti, Synlett, 1997, 93; J.J.kiddle, Tetrahedron Lett, 2000, **41** 1339).

$$Ph_3P \quad XCH_2R_1 \xrightarrow{MW} Ph_3\overset{+}{P}-CH_2R_1X^- \xrightarrow[MW]{base}$$

Triphenyl phosphine — Organic halides — Phosphonium salt

$$\longrightarrow \left[Ph_3\overset{+}{P}-\overset{-}{C}HR_1 \longleftrightarrow Ph_3P=CHR_1 \right] \xrightarrow[MW]{R_2C=O} R_1-CH=CR_2$$

ylide — Phosphorane — Olefin

7.20 CONCLUSION

Microwave technique has been used for greener alternatives to synthetic organic transformations. It leads to substantial savings in time and energy for many organic transformations. The technique provides rapid and relatively inexpensive access to very high temperatures and pressures. Transformations using microwaves can be conducted in water, in organic solvents and without the need of solvent in the solid state. The latter approach in beneficial as the reccovery of solvents from conventional reaction systems always result in some loss. Using this technique, a number of procedures involving acetylation, deacetylation, deprotection of benzyl ester and t-butyldimethyl silyl esters have been developed. The technique is useful for saponicification, dethicacetalization of thioacetals and thioketals of aldehydes and ketones besides decarboxlation of aromatic carboxylic acids. This technique has also been used for conductiong Cannizzaro reaction, Diels-Alder reaction, Fisher-Indole synthesis and Modelung Indole synthesis. All these aplication find use in synthetic organic transformations.

Chapter 8

Enzymatic Transformations

8.1 BENZOIN

Benzoin is generally obtained from benzaldehyde using cyanide iron catalysit (See Section 2.4)

An enzymatic green synthesis of benzoin is by the reaction of benzaldehyde with a biological catalyst, thiamine hydrochloride (R.Breslow, J.Am. Chem. Soc., 1958, **80**, 3712).

$$\underset{\text{Benzaldehyde}}{2C_6H_5CHO} \xrightarrow{\text{Thiamine}} \underset{\text{Benzoin}}{C_6H_5\overset{\overset{\displaystyle O}{\|}}{C}-\overset{\overset{\displaystyle OH}{|}}{CH}-C_6H_5}$$

Various steps involved in the mechanism are given below:

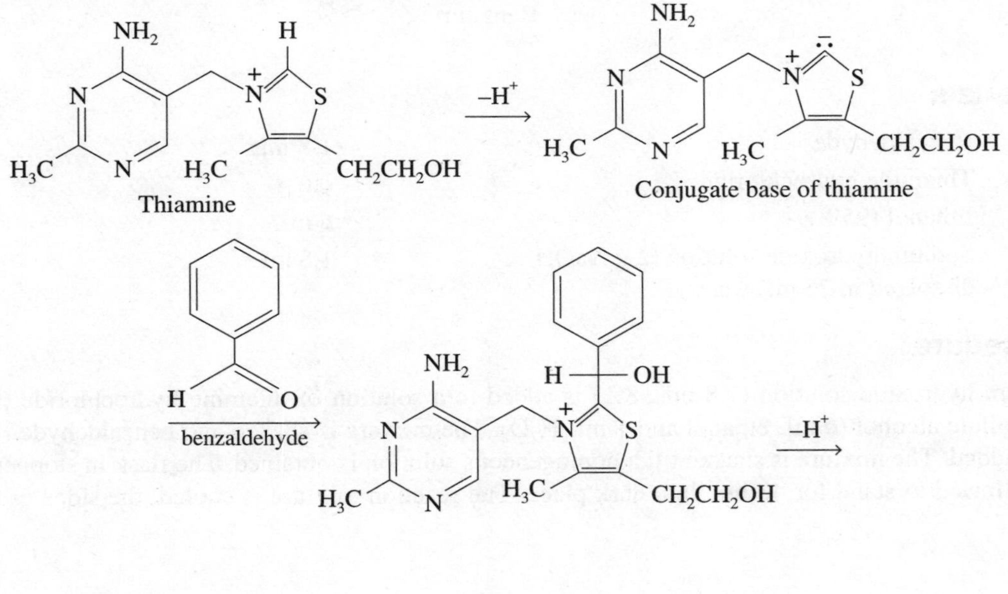

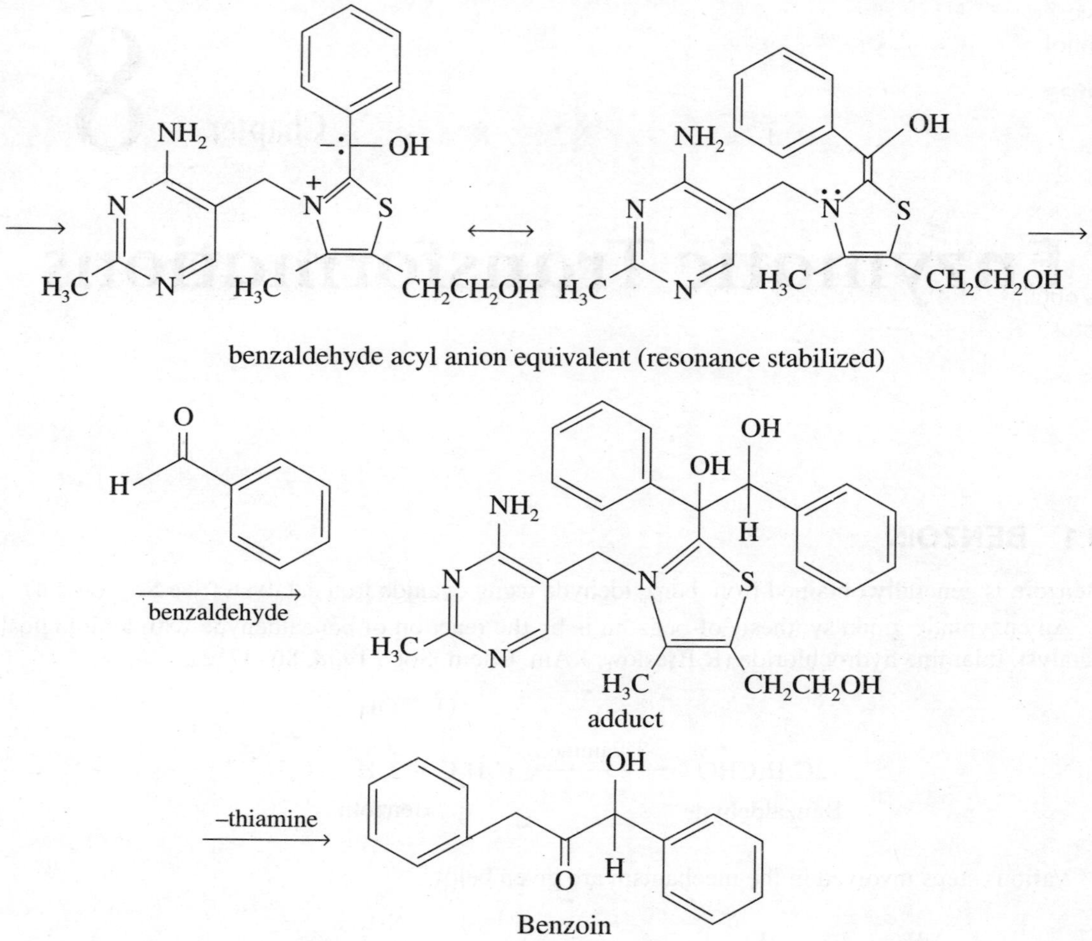

benzaldehyde acyl anion equivalent (resonance stabilized)

benzaldehyde

adduct

–thiamine

Benzoin

Materials

Benzaldehyde	1.8 mL
Thiamine hydrochloride	0.6 g
Ethanol (95%)	6 mL
Sodium hydroxide solution (2 g NaOH dissolved in 25 mL water)	1.8 mL

Procedure

Sodium hydroxide solution (1.8 mL, 8%) is added to a solution of thiamine hydrochloride (0.6 g) in dilute alcohol (6 mL ethanol and 1 mL H_2O). The mixture is shaken and benzaldehyde (1.8 mL) added. The mixture is shaken till a homogeneous solution is obtained. The flask in stoppered and allowed to stand for 48 hrs. in a dark place. The reaction mixture is cooled, the sides of the flask scratched (to induce crystallization) and the separated benzoin filtered. It is crystallized from alcohol, m.p. 134-35°. Record the yield.

Notes

1. Pure benzaldehyde and pure thiamine hydrochloride is used.
2. See also preparation of benzion using cyanide as catalyst (Section 2.4)

8.2 ETHANOL

It is obtained by the fermentation of sucrose by invertase, which converts sucrose into glucose and fructose, followed by zymase, which converts glucose and fructose into ethanol

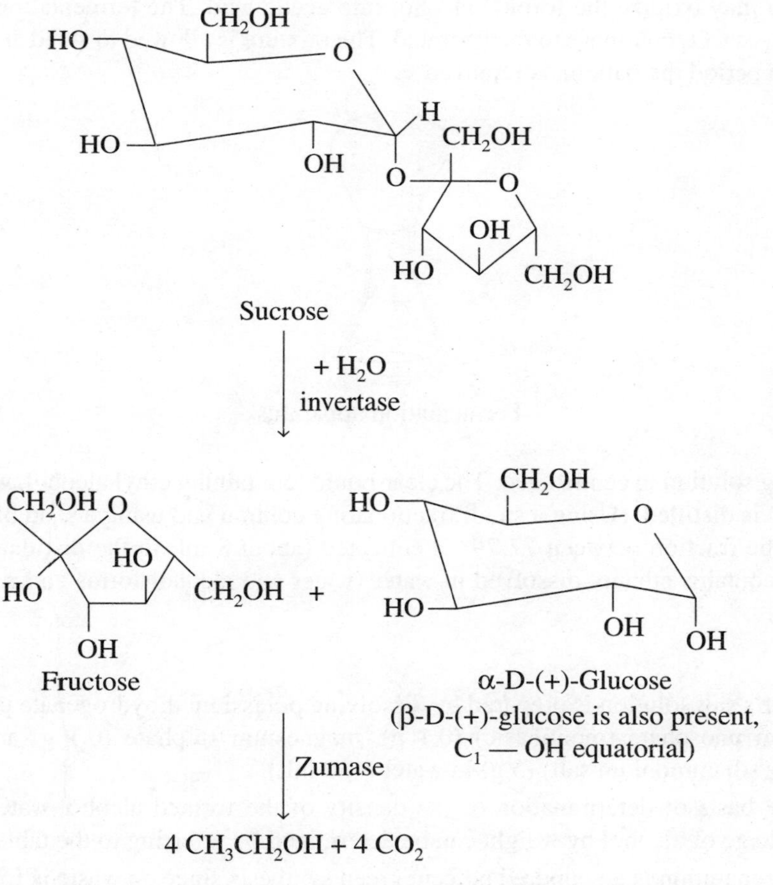

Sucrose

$+ H_2O$
invertase

Fructose + α-D-(+)-Glucose
(β-D-(+)-glucose is also present, C_1 — OH equatorial)

Zumase

$$4\,CH_3CH_2OH + 4\,CO_2$$

The above process is also used for the preparation at alcohol on industrial scale. However, on industrial scale, molasse's (the mother liquid left. after the isolation of sucrose) is used in place of sucrose. By this process it is not possible to obtain more than 10-15% ethanol, since fermentation is inhibited if the concentration of ethanol exceeds 15%. More concentrated alcohol is obtained by fractional distillation.

Materials

Sucrose	20 g
Bakers yeast	2 g
Pasteur's Salt Solution	20 mL

Procedure

Sucrose (20 g) is placed in a Erlenmeyer flask (1000 mL capacity). Water (140 mL) warmed to 25-30°, pasteur's salt solution (20 mL) and Bakers yeast (2 g) is added. The mixture is shaken so as to thoroughly mix. Attach a balloon directly to the flask (as shown in the fig. below). The balloon expands by the evolved gas as fermentation proceeds. This procedure is helpful to exclude contact with air, which may oxidize the formed alcohol into acetic acid. The fermentation is allowed to proceed as long as CO_2 continues to the liberated. The mixture is allowed to stand at 30-35° for 5-6 days. After this period the balloon is removed

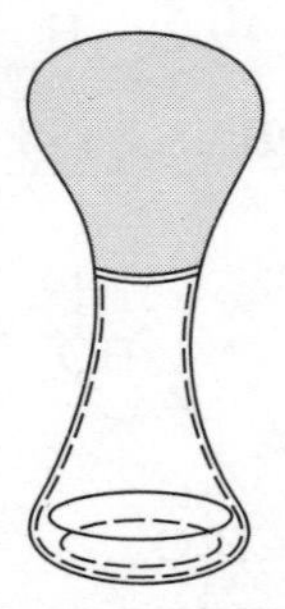

Fermentation apparatus

The resulting solution in centrifuged. The clear liquid (containing ethyl alcohol, water and some dissolved salts) is distilled. (Using a small fractionating column and using a sand bath (150-200°) for heating). The fraction between 77-79° is collected (about 8 mL in the distillate is obtained). This distillate contains ethanol dissolved in water (water and alcohol forms an azeotrope, which boils at 78°).

Notes

1. Pasteur's salt solution is prepared by dissolving potassium dihydrogenate phosphate (1 g) calcium phosphate (monobasic) (0.1 g), magnesium sulphate (0.1 g) and ammonium tartrate (diammonium salt) (5 g) in water (430 mL)
2. On the basis of determination of the density of the formed alcohol-water mixture, the percentage of alcohol by weight can be determined by referring to the tables.
3. This preparation is an hundred percent green synthesis since no waste is formed.

8.3 (S) –(+)–ETHYL 3-HYDROXYBUTANOATE

It is obtained by the reduction ethyl acetoacetate using bakers yeast.

$$CH_3COCH_2COOC_2H_5 \xrightarrow[\text{H}_2\text{O, sucrose, 20-30°C}]{\text{baker's yeast}} CH_3CH(OH)CH_2COOC_2H_5$$

Ethyl acetoacetate → (S)-(+)–Ethyl 3-hydroxybutanoate

Materials

Ethyl acetoacetate	9 g
Sucrose	134 g
Dry Bakers yeast	15.6 g

Procedure

Water (225 mL), sucrose (67 g) and bakers yeast (7.8 g) is taken in a conical flask (1 litre capacity). The solution is stirred with a magnetic needle. The apparatus is fitted up as shown in the diagram. Ethyl acetoacetate (9 g) is added to the flask and the fermentation mixture is kept at room temperature and stirring continued for 24 hr. Sucrose solution (67 g in 300 mL water) is kept at 40° and added to the fermenating liquid along with bakers yeast (7.8 g). The mixture is allowed to stand for 48 hr.

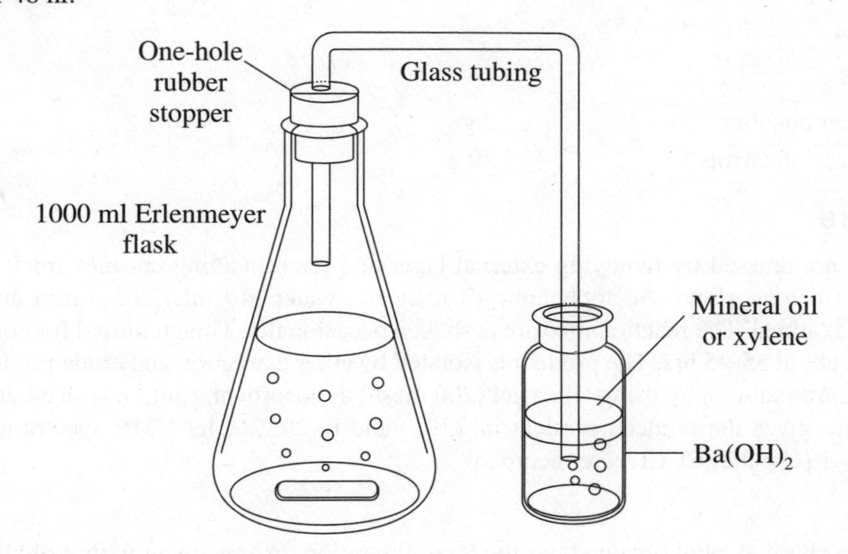

After the reaction is complete, the reaction mixture is filtered by decantation using a bluchner funnel (filter paper) containing filter acid. The filtrate is extracted with ether (3 × 50 mL), ether extract dried ($MgSO_4$) and distilled in nitrogen atmosphere to give the reduced product (7-8 mL). The crude product is purified by column chromatography using alumina as adsorbent and eluting with mehtylene chloride. Distillation of the solvent from the eluate give the pure hydroxyl ester.

Notes

1. Record the 1 R spectrum of the hydroxy ester.
2. This is a green synthesis since no byproduct which may cause pollution are formed
3. Reduction ethyl β-ketovalrate (O, CO_2Et) with Bakers yeast gives the corresponding (R)-alcohol, (OH, CO_2Et).

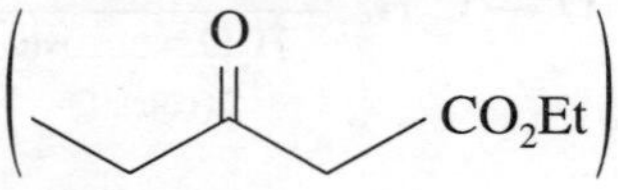

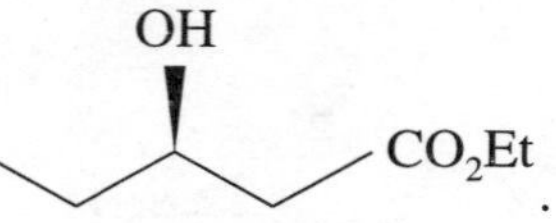

8.4 1-PHENYL-(1S) ETHAN–1–OL

It was earlier obtained by the reduction of acetophenone by Baker's yeast (O.P. Ward and C.S. Young, Enzyme Microbial Technol, 1998, **12**, 4822). However the isolation of the product is not straight forward. It has now been obtained by a convenient procedure involving reduction of acetophenone with Daucus carrota roots.

$$C_6H_5C - \overset{\overset{O}{\|}}{C} - CH_3H_3 \xrightarrow{D \cdot carota} C_6H_5\overset{\overset{OH}{|}}{C}H - CH_3$$

Acetophenane → 1-Phenyl-(IS) ethan-1-ol (73%)

Materials

Acetophenone	2 g
Slices of carrots	20 g

Procedure

The carrots are dressed by removing external layer and the remaining cut into small thin pieces (approx. 1 cm long slice). Acetophenone (2 mL) and water (40 mL) are stirred and slices of carrots (20 g) added. The reaction mixture is shaken occasionally. Time required for completion of reduction is about 35-45 hrs. The product is isolated by ether extraction and crude product purified by column chromatography using silica gel (200 mesh) as adsorbent. Elution with ether-petroleum ether mixture gives the reduced product in 73% yield (ee 92%). Its NMR spectra and rotation $[\alpha]_D^{25} = -39.1$ (C = 3.5, $CCCl_3$) are recorded.

Notes

1. The chiral alcohol obtained has the S configuration, in agreement with prologs rule.
2. Using the above procedure, a number of ketones (like p-Cl, p-Br, p-F, p-NO_2, p-Me, p-OMe, p-OH) can be conveniently reduced to the correspondings alcohol with 90-98% ee and in good yields (70-80%).

3. This procedure can also be used for the reduction of β-ketoesters, cyclic ketones, azido ketones and even open chain ketones like 2-butanone, 2-pentanone etc.

8.5 2-(QUINOLIN—8-YLOXY)-NAPHTHO [2, 1-b]PYRAN-3-ONE

It is obtained by cyclocondensation of 2-hydoxy-1-naphthaldehyde with quinoline-8-yloxy acetic acid using microwaves in the presence of enzyme **lipase** form pseudomonas speies in presence of DBU.

HO, OHC, N, $HOOCCH_2C$—O, +

Quinolin-8-yloxy acetic acid 2-Hydroxy-1-napthaldehyde

Lypase, DBU, DMF / Microwave, 40° / 4-6 min

2-(Quinolin-8-yloxy)-naphtho [2,1-b]pyran-3-one (68%)

Procedure

A mixture of 2-hydrxy-1-naphthaldehyde (5 m mole),quinolin-8-yloxyacetic acid (6.25 m mole), 1, 8-diazobiscyclo [5.4.0]-undecene-7 (DBU) (1 m mole) in DMF (15 mL) and Lipase (0.5 g) is irradiated in a MW oven at 40° for 4-5 min. The residual solid is filtered and washed with water to give the required 2-(quinolin-8-yloxy)-naphtho [2, 1-b] pyran-3 one, m.p. 179° in 68% yield.

Notes

1. M. Kidawai, S. Kohli, R.K.Saxena, R. Gupta and S. Bardoo, Ind. J. chem., 1998, 963
2. The required quinolin-8-yloxyacetic acid is obtained by treatment of 8-hydroxyqunoline with ethyl bromoacetate followed by hydrolysis of the ester.
3. Using the above procedure a number of other 2-substitued naphthol [2, 1-b] pyram -3 ones were obtained.

8.6 REDUCTION OF ALDEHYLES AND KETOSES

Enzymatic reduction of crotonaldehyde (trans isomer) with Beauveria sulfurescens give 2-buten-1 ol in 80 % yield (M. Desart, A. Kergemard, M.F. Record and H. Vesclambre, Tetrahetron, 1981, **37**, 3825)

$$CH_3CH{=}CHCHO \xrightarrow{\text{B. Suflurescens}} CH_3CH{=}CHCH_2OH$$

Crotanaldehyde 2-Buten-1 ol 80%

However, reduction of 2-methyl-2-pentenal on reduction with B. Sulfurencens gives a mixture of 31% of the conjugated alcohol and 69% of the completely reduced product.

2-Methyl-2-pentenal (CHO) $\xrightarrow{\text{B. sulfurescens}}$ CH_2OH + CH_2OH

2-Methyl-2-pentenol (31%) 2-methyl-1-pentanol (69%)

Smaller ketones like 2-butanones on reduction with Thermoanaerobium brockii gives the (R) alcohol. However, larger ketenes like 2-hexanones on reduction with Thermoanaerobium brockii gives the (S) alcohol. (E.Kienan, E.K. Hateli, K.selh and R. Lamed, J. Am.Chem. Soc., 1986, **108**, 162)

2-Butanone $\xrightarrow{\text{T. Brockii}}$ (R)-Alcohol (12%, 40%ee)

2-Hexanane $\xrightarrow{\text{T. Brockii}}$ (S)-Alcohol (85%, 96% ee)

Ketones can also be reduced to the corresponding (S) alcohol by using Daucus carota roots (see preparation of 1-phenyl-(1S) ethan-1-ol, note 2, Section 8.4).

8.7 MISCALLANLOUS ENZYMATIC TRANSFORMATIONS

1. Oxidations

Enzymes are used in a number of oxidations. Some examples are given below

(a) Oxidation at ethyl alcohol to acetic acid

Ethyl alcohol on oxidation with bacterium acetic in presence air gives acetic acid. In fact, this is a commercial method for the manufacture of acetic acid. The process is known as **quick-vinegar process**.

$$CH_3CH_2OH + O_2 \xrightarrow{\text{Bacterium acetic}} CH_3COOH + H_2O$$

Ethylalcohol Acetic acid

(b) Baeyer-villiger reaction

This can be affected using the enzyme cyclohexanone oxygenase from Acinetobacter strain NCIB 9871. Thus, cyclohexanone is convicted into ε-caprolactone (J.M.Schwab, W.-b, Li and L.P. Thomas, J. Am. Chem. Soc., 1983, **10**, 4800)

$$\text{Cyclohexanane} \xrightarrow[\text{glucose 6-phosphate}]{\text{Cyclohexanone oxygenase, NADPH, } O_2} \varepsilon\text{-Caprolactone (28\%)}$$

Similarly, phenyl acetone is converted into **benzyl acetate** (B.P Branchaud and C.T. Walsh, J. Am. Chem,. Soc., 1985, **107**, 2153).

$$\underset{\text{Phenylacetone}}{C_6H_5CH_2COCH_3} \xrightarrow[\text{Enz-FAD, NADPH, } H^+]{\text{Cyclohexanone oxygenase, NADPH, } O_2} \underset{\text{Benzylacetate}}{C_6H_5CH_2OCOCH_3}$$

Most of the **Balyer-Villiger Oxidations** of steroids are accomplished biochemically. For details see C.Tamn, A. Gubler, G. Juhasz, E.Weiss-Berg and W.Ziircher, Helv. Chim. Acta, 1963, **46**, 889; R.L. Prairie and P.Talay Biochemistry, 1963, **2**, 203.

(c) Oxidation benzene to trans, trans muconic acid

Benzene can be oxidized by Micrococcus speroids-like organisms into trans, trans-muconic acid (A.Kleinzeller and Z Fencl, Chem. listy, 1952, **46**, 300; Chem. Abstr., 1953, 47, 4290).

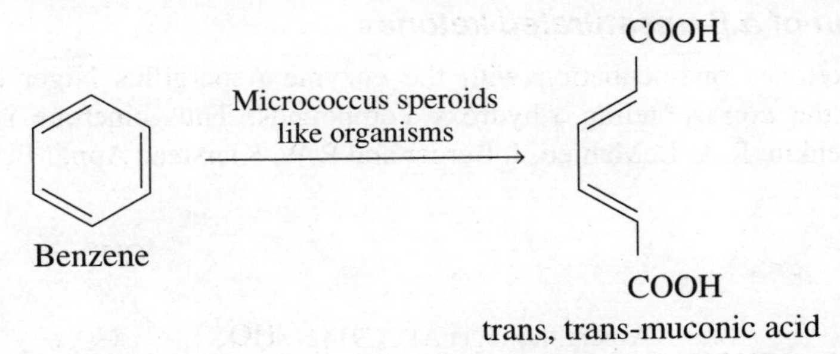

(d) Oxidation at Oleic acid

Oxidation of oleic acid with pseudomons sp. gives 11-hydroxystearic acid (L.L.Wallen, R.Y.Benedict and R.G.Jackson, Arch. Biochem. Biophys. 1962, **99**, 249)

$$CH_3(CH_2)_7CH = CH(CH_2)_7COOH \xrightarrow{\text{Pseudomonas sp.}} CH_3(CH_2)_7CH(OH)CH_2(CH_2)_7COOH$$

Oleic acid → 10-Hydroxystearic acid

(e) Oxidation of primary alkyl halides to carboxylic acids

Long chain aliphatic halides on incubation with the yeast Torulopsis gropengiesseri gives dicarboxylic acids. The enzymatic oxidation starts at the terminal methyl groups and gives ω-halogeno alkanoic acids, which are finally converted into α, ω-dicarboxlic acids. (D.F. Jones and R.Home, J. Chem. Soc. 1968, 2816; and 2801)

$$CH_3(CH_2)_{16}CH_2X \xrightarrow[\text{gropengiesseri}]{\text{Torulopsis}} [HO_2C(CH_2)_{16}CH_2X + HOOC(CH_2)_{14}CH_2X] \rightarrow HOOC(CH_2)_{16}COOH + HOOC(CH_2)_{14}COOH$$

X = F, Cl, Br, I

X	$HOOC(CH_2)_{16}COOH$	$HOOC(CH_2)_{14}COOH$
X = F	98	4.5%
Cl	18%	29%
Br	17%	29%
I	15%	30%

In the above enzymatic oxidation, dicarboxylic acids with two less carbons are concomitantly produced.

(f) Hydroxylation of α,β-unsaturated ketones

α,β Unsaturated ketones on incubation with the enzyme Aspergillus Niger or Streptomyces aureofaciens give the corresponding γ-hydroxy compounds. Thus cinerone is converted into cinerolone (B. Tebenkin, R.A. LeMahieu, J. Berger and R.W. Kierstead App. Microbiol, 1969, **17**, 714).

Cinerone (3-methyl-2-($CH_2CH{=}CHCH_3$)cyclopent-2-en-1-one) $\xrightarrow[\text{OR Streptomyces aureofaciens ATCC 1076L 48-90 hr.}]{\text{Aspergillus niger ATCC9142}}$ Cinerolone (45-60%) (HO, CH_3, $CH_2CH{=}CHCH_3$ substituted cyclopentene)

(g) Oxidation of aldehydes to carboxylic acids

The biochemical oxidation of phenylacetaldehyde with cyclohexanone oxygenase yields phenylacetic acid along with smaller amounts of benzyl formate and benzyl alcohol (B.Branchand and C.T.Walsh, J.Am.Chem. Soc., 1985, **107**, 2153)

$$C_6H_5CH_2CHO \xrightarrow{\text{Cyclohexanone oxygenase}} C_6H_5CH_2COOH + C_6H_5CH_2OCHO + C_6H_5CH_2OH$$

Phenylacetaldehyde → Phenylacetic acid (65%) + Benzylformate (12%) + Benzyl alcohol (23%)

The conversion of aldehydes into carboxylic acid can normally be achieved by **cannizzaro reaction**. However, the yields are only 50% (The alcohol, C_6H_5-CH_2OH is also obtained in 50% yield).

Plyenyacetic acid can also be obtained by enzymatic oxidation ethyl benzene, butyl benzene or dodecylbenzene (see the flowing sub section)

(h) Oxidation of aromatic side charins to carboxyl group

The aromatic side chains on oxidation with the enzyme. **Nocardia strain** gives phenyl acetic acid (J.B. Davis and R.L. Raymond, Appl. Microbiol, 1961, **9**, 38). Some examples are given below.

$$C_6H_5CH_2R \xrightarrow[30°]{\text{Norcardia strain 107-332}} C_6H_5CH_2COOH$$

Ethyl benzene $R = CH_3$ → Phenyl acetic acid

n-Butyl benzene, $R = n\text{-}CH_2CH_2CH_3$

Dodecyl benzene, $R = CH_2C_{10}H_{21}$

(i) Oxidation of primary amines to carboxylic acids

Primary amine (like tryptamine) can be oxidized biochemically using **Hygrophorus concius** to the corresponding carboxylic acid (indolone-3-acetic acid) (D.J.Siehr, J.Am.chem.Soc., 1961, **83**, 2401)

Tryptamine (3-($CH_2CH_2NH_2$)indole) $\xrightarrow[27°, 7 \text{ days}]{\text{Hygrophorus conicus}}$ 2-Indolone-3-acetic acid (43%) (3-(CH_2COOH)-2-indolone)

(j) Oxidation of primaryamines to nitro compounds

Primary amines can oxidized to the corresponding nitro compound by the enzyme **streptomyces thiolutes** (S. Kawai, T.Oshima and F.Egami; Beochem. Biophys, Acta, 1965, **97**, 391)

H_2N—C6H4—COOH —(Streptomyces thioluteus, pH7, 30°)→ O_2N—C6H4—COOH

p-Amino benzoic acid → p-Nitrobenzoic acid 90%

(k) Conversion of Penicillin into 6-APA

One of the most important biocatalytic conversion is the conversion of Penicillin's into 6-APA by the enzyme **'Penacylase'**

Penicillin G —(Pencylase, H_2O, 37°)→ 6-APA

(l) Miscellaneous applications of enzymes

(a) Glycosidase are used in large quantity for converting corn starch to glucose (C.Gruesbeck and H.F. Rase, Ind. Eng. Chem. Proc, Dev., 1972, 11, 74) and glucose isomerase catalyses the equilibration of glucose and fructose (H.H. Weetall, Process Biochem, 1975, 10, 3)

(b) Fumaric acid is converted into aspartic acid by the addition of ammonia catalysed by aspartase (T.Tosa, T.Sato, T.Mori, y. Matuo and I. Chibata, Biotechnol. Biology, 1973, **15**, 69).

(c) L-Citraline and L-arginine have been prepared on a larger scale using hydrolytic deamination (Y. Izumi, I-clibata and T.Itoh, Ang. Chem.Int.Ed.Engl., 1978, **17**, 176.

(m) Reductions

Ethylacetoacetate —(Bakers yeast)→ (S) Ethy 3-hydroxy butyrate (67)%

Ethy β-ketovalerate —(Bakers yeast)→ (R) Ethyl β-hydroxy valerate

Beauveria sulfurencens

CHO → CH_2OH

trans crotonaldehyde → 2-Buten-1-ol 80%

2-Butanone —T. Brockii→ (R)- Alcohol (12%)

2-Hexanone —T. Brockii→ (S)-Alcohol (85%)

Besides the enzymatic transfermation given above a large number other transformation inducing synthesis of important products is also possible. For details see V.K.Ahluwalia, Enzymes for Green Organic Synthesis, Narosa Puplishing House, 2010 and the references cited therein.

8.8 CONCLUSION

A large number of important transformations in the context of green chemistry are with the help of enzymes. Most of the enzymatic transformations are conducted in aqueous medium at ambient temperature. Most of the biocayalytic transformations are one step process. The conversions are sterospecific.

Chapter 9

Transformations in Ionic Liquids

9.1 1-ACETYLNAPHTHALENE

It is obtained by **Friedel-crafts reaction** of naphthalene with acetyl chloride in presence of ionic liquid [emin] Cl-$AlCl_3$ [emin = 1 – methyl-3-ethylimidazolium cation]

Napthalene + CH_3COCl $\xrightarrow[\text{r.t}]{\text{[emin] Cl-AlCl}_3}$ 1-Acetyl napthalene ($COCH_3$)

Normally, Friedel-crafts acetylation of naphthalene gives the thermodynamically favourable 2-isomer. However, in ionic liquid the thermodynamically unfavourable 1-isomer is obtained.

Materials

Naphthalene	10.1 g
Acetyl chloride	6 g
[emin] Cl-$AlCl_3$	20 g

The required [emin] Cl is prepared by the procedure described by R.S. Varma and V.V. Namboodri, Chem. Commun., 2001, 643. The procedure involves mixing ethyl chloride (7.8 g) and (1-methyl-3-ethylimidazole (11.2 g) and heating the mixture in unmodified house hold microwave oven (240 W) for 1 min. till a clear single phase is obtained. The resultant ionic liquid is cooled, washed with ether (3 × 20 mL) (to remove the unreacted starting material).

Procedure[1]

A mixture of naphthalene (10.1 g), acetyl chloride (6 g) and the ionic liquid [emin] Cl-$AlCl_3$ (20g) [obtained by mixing equimolar amounts of 1-methyl-3-ethylimidazolium chloride with anhydrous aluminium chloride] is kept for 1 hr. The mixture is extracted with ether and the residual product is crystallized form alcohol to give 1-acetyl naphthalene in 90% yield. Its NMR spectra is recorded.

1. C.J. Adams, M.J. Earle, G. Roberts and R. Seddon, Chem. Commun., 1998, 2097.
2. The recovered ionic liquid can be used again

9.2 4-ACETYL-1-METHYLCYCLOHEXENE

It is prepared by **Diels-Alder reaction** of isoprene (diene) and but-2-en-3-one (dienophile) in ionic liquid, [b min] [BF_4] at 20° for 2 hr.

$H_3C—C(=CH_2)—HC=CH_2$ (Isoprene) + $CH_2=CHCOCH_3$ (But-3-en-2-one) $\xrightarrow[20°,\ 2\ hr]{[emin]\ BF_4}$ 4-Acetyl-1-methyl cyclohexene

The required 1-ethyl-3-methylimidazolium tetrafluoroborate, [emin] BF_4 is prepared by the procedure described in literature (J.D. Holbery and K.R. Seddon, J.Chem. Soc. Dalton Trans, 1999, 2133). The procedure involved slow addition of tetrafluoroboric acid (12.2 mL, 0.116 mol 48% solution in water) to a rapidly stirred slurry of silver (I) oxide (13.49 g, 0.058 mol) in water (50 mL). The reaction mixture is covered with aluminium foil in order to prevent photodegradation. The stirring is continued (1 hr) till all Ag(I) oxide has completely reacted and a clear solution results. A solution of 1-ethyl-3-methylimidazolium bromide (22. 24 g. 0116 mol) in water (200 mL) is added to the reaction mixture and stirring continued for 2 hrs. The separated silver bromide is removed by filtration. The filtrate is heated at 70° under reduced pressure to remove the solvent to give the ionic liquid [emin] BF_4 as pale yellow liquid, yield 93%.

Finally, isoprene, but-3-en-2-one and [b min] BF_4 [in the molar ration 1.5 : 1.0 : 1.0] was kept at 20° for 2 hr. to give 4-acetyl-1-methyl cyclohexene in 90% yield (endo : exo. ratio 4:0 :1). The final product is isolated by extraction with ether.

Notes

1. The procedure described is that of M.J. Earle, P.B. Mc Cormac and H.R. Seddon, Green Chem., 1999, 23
2. Using isoprene (diene) and ethyl acrylate in ionic liquid [b min] BF_4 at 70° for 24 hr. yielded ethyl 4-methyl-3-cyclohexene carboxylate in 93% yield (endo: exo. 4:1)

$$H_3C-C(=CH_2)-CH=CH_2 \text{ (Isoprene)} + CH_2=CHCO_2Et \text{ (Ethylacrylate)} \xrightarrow{[emin]\,BF_4} \text{4-Methyl-3-cyclohexnene carboxylate 93\%}$$

3. The ionic liquid can be reused.

9.3 p-METHYLACETOPHENONE

It is obtained by **Friedel-crafts reaction** of toluene with acetyl chloride in the presence of ionic liquid, [emin]Cl-$AlCl_3$

$$\text{Toluene} + CH_3COCl \xrightarrow[\text{R.T.}]{[emin]\,Cl-AlCl_3} \text{p-Methylacetophenone}$$

Materials

Toluene (dry)	42 mL
Acetyl chloride	50 mL
[emin] Cl-$AlCl_3$	40 g

Procedure

The procedure is similar to that used for the preparation of 1-acetylnaphthalene. p-Methyl acetophenone (b.b. 225°) is obtained is 95% yield compared to about 55% by the routine Friedel-Craftes reaction

Notes

1. Using the above procedure chlorobenzene and anisole give the corresponding 4-acetyl compound in 97-99% yields.
2. The above procedure can also be used for the preparation of acetophenone and 2-hydroxy-4-methoxy acetophenone

3. Ionic liquids can also be used for **Friedel-crafts alkylation**. Thus the reaction of benzene with tertiary butyl chloride or chloroform in ionic liquid give tertiarybutyl benzene, $C_6H_5C(CH_3)_3$ (b.p. 168.5°) and triphenylmethane (Ph_3CH) (m.p. 92°) respectively in more than 80% yields.

9.4 OXIDATION OF BENZYLIC ALCOHOLS TO ALDEHYDES.

The oxidation of benzyl alcohol with $KMnO_4$ in 1-butyl-3-methylimidazolium tetrafluoroborate, [bmin] [BF_4] ionic liquid at room temperature gave benzaldehyde in 90% yield in 1 hr. (A. Kumar, N. Jain and S.M.S. Chauhan, Synth. Commun., 2004, **34**, 2835)

$$\underset{\text{Benzyl alcohol}}{C_6H_5CH_2OH} \xrightarrow[\text{[bmin] [BF}_4\text{] 1 hr. RT}]{KMnO_4} \underset{90\%}{C_6H_5CHO}$$

Using this procedure a number of substituted benzyl alcohols could be oxidised to the corresponding aldehydes.

9.5 MISCELLONEOUS APPLICATIONS OF IONIC LIQUIDS

(i) Alkylation of Active Methylene Compounds

This is an important reaction for C-C bond formation. It is carried out in the ionic liquid, N-butylpyridinium tetrafluoroborate, [bpy] [BF_4], which in a recyclable solvent. Thus, alkylation of Meldrum's acids with a number of alkyl halides at 60-70° in presence of triethylamine gives exclusively dialkylated products. (C. Su, Z.-C. Chen and Q. G. Zhen Synth. Commun., 2003, **33**, 2817)

$$\underset{\text{Meldrum's acid}}{\text{Meldrum's acid}} + RCH_2X \xrightarrow[60\text{-}70°]{Et_3N,\ [bpy][BF_4]} \text{5,5-bis}(CH_2R)\text{ Meldrum's acid}$$

(ii) Cyclopropanation

Alkenes can be enantioselectively cyclopropanated with ethyl diazoacetate using a bis(oxazoline)-copper catalytic system immobilised in an ionic liquid like 1-ethyl-3-methylimidazolium tetrafuoroborate, [emin] [BF_4] (J.M. Fraile, J.I. Garcicr, C.I. Herrierias, J.A. Mayoral, D.Carrie and M.Vaultier, Tetrahedron Assymetry, 2001, **12**, 1891). The catalyst system is recyclable and isolation of final product is simple.

Styrene + N_2CHCO_2Et (Ethyl diazoacetate) $\xrightarrow[\text{[emin] [BF}_4\text{] or [emin] [NTf}_2\text{]}]{\text{Cu (II)L}}$ (cyclopropane products) + (cyclopropane products)

L = (bis-oxazoline ligand: O, N, N, O, R, R) R = Ph, ${}^tBu$

(iii) Synthesis of Ethers

A convenient and efficient procedure for the synthesis of diaryl ethers by the reaction of phenols and aryl halides in the presence of a base catalysed by CuCl immobilised in the ionic liquid, [bmin] [BF_4] is described (S.M.S. Chauhan, N. Jain, N. Kumar and K.A. Srinivas, Synth. Commun., 2003, **33**, 3607). The yields of diaryl etters in much higher then those obtained in conventional solvents like DMF. In fact, this is a convenient method for carbon-oxygen bond formation.

Phenols (R–C_6H_4–OH) + Aryl halides (Cl–C_6H_4–R′) $\xrightarrow[\text{[bmin] [BF}_4\text{]}]{\text{CuCl, base}}$ Ethers (R′–C_6H_4–O–C_6H_4–R′)

(iv) Oxidations

A number of transition metal-catalysed oxidation reaction have been performed in low melting imidazolium and pyridinium ionic liquids. A typical example is the oxidation of ethylbenzene to phenylacetaldehyde using bis(acetylacetonato) nickel (II) immobilised in ionic liquid [bmin] [PF_6] at atmospheric presume (C.R. Alcantara, L. Canoira, P. Guilhermo –Joao and J.P. Perez-Mendo, Appl. catal, A, 2001, 218, 269)

$$\underset{\text{Ethyl benzene}}{C_6H_5CH_2CH_3} \xrightarrow[\text{[bmin] [PF}_6\text{]}]{\text{Ni (acac)}_2\text{, O}_2} \underset{\text{Phenylacetaldehyde}}{C_6H_5CH_2CHO}$$

Aromatic aldehydes are oxidised to the corresponding carboxylic acids using bis(acetylacetonato) nickel (II) immobilized in ionic liquid, [bmin] [BF_6] and oxygen at atmospheric pressure (R. Alcantara, L. Canoirae, P. Guilhermo-Joao and J.P. Perez-Mendo, Appl. Catal. A, 2001, **218**, 269)

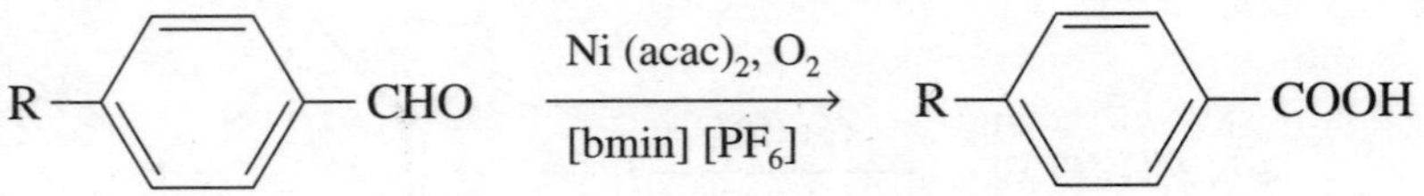

$$R-C_6H_4-CHO \xrightarrow[\text{[bmin] [PF}_6\text{]}]{\text{Ni (acac)}_2\text{, O}_2} R-C_6H_4-COOH$$

Alcohols can be oxidised to the corresponding aldehydes and ketones in dry ionic liquids using $Pd(OAc)_2$ as catalyst and O_2 as an oxidant (K.R. Seddon and A. Stark, Green Chem. 2002, **4**, 119)

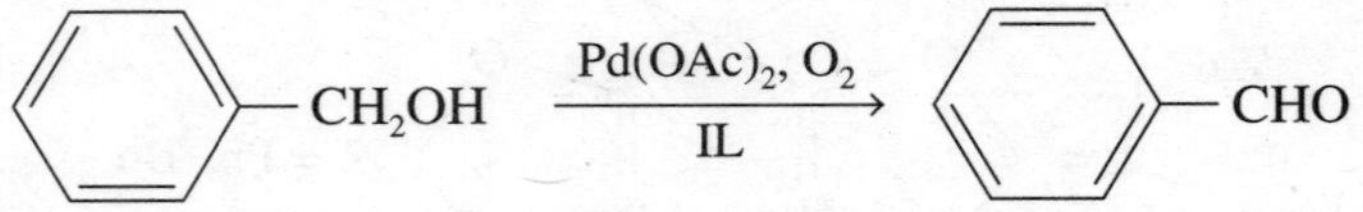

$$C_6H_5-CH_2OH \xrightarrow[\text{IL}]{\text{Pd(OAc)}_2\text{, O}_2} C_6H_5-CHO$$

Benzyl alcohol Benzaldehyde

Oximes can be chemoselectively oxidised with H_2O_2 catalysed by phosphotungstic acid in ionic liquid at room temperature generating the corresponding starting carbonyl compounds in excellent yield (N. Jain, S. Kumar and S.M.S. Chauhan, Tetrahedron, 2005, **61 (5)**, 1015)

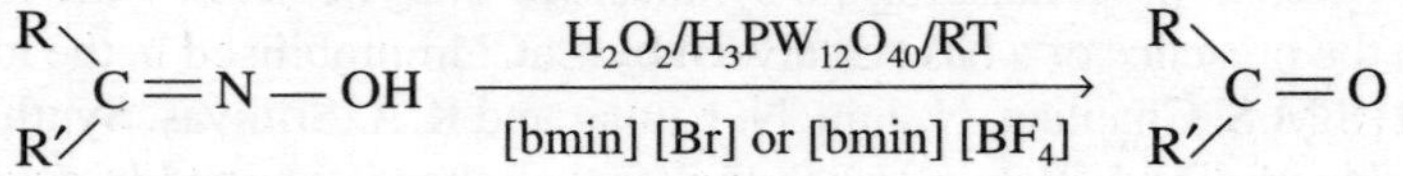

$$RR'C=N-OH \xrightarrow[\text{[bmin] [Br] or [bmin] [BF}_4\text{]}]{H_2O_2/H_3PW_{12}O_{40}/RT} RR'C=O$$

Alkenes and allylic alcohols could be oxidised to the corresponding epoxides in excellent yield using methyltrioxorhenium (MTO) as catalyst and urea-hydrogen peroxide (UPH) as the oxidant and the ionic liquid [bmin] [PF_6], in which both the catalyst and oxidant are completely soluble (G.S. Owens and M.M. Abu-omar, Chem. Commun., 2000,

Olefins can also be epoxidised with NaOCl using Jacobsen's Chiral Mn (II) Salen, imbobilized in ionic liquid [bmin] [PF_6] (C.E. Song and E.J. Roh, Chem. Commun., 2000, 837). An example is given below.

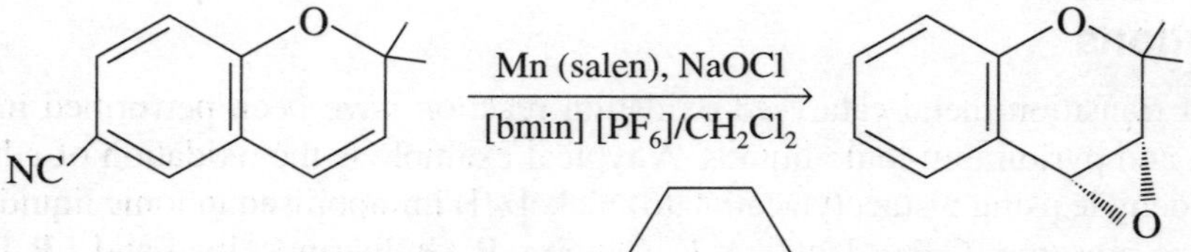

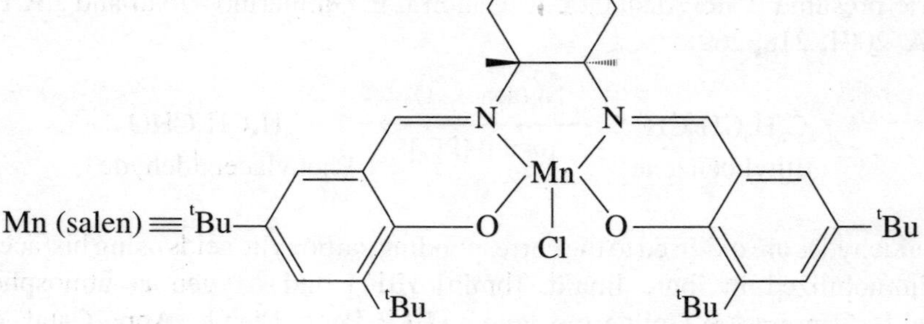

$$\xrightarrow[\text{[bmin] [PF}_6\text{]/CH}_2\text{Cl}_2]{\text{Mn (salen), NaOCl}}$$

NC O O O

Mn (salen) ≡ ᵗBu ᵗBu ᵗBu ᵗBu N N Mn O O Cl

Asymmetric dihydroxylation of olefins is normally carried out using osmium catalysts. However, the high cost, toxicity and contamination of the product with osmium catalyst restricts its use in industry. This problem is solved by using ionic liquids. The dihydroxylation is carried out either in biphasic [bmin] [PF_6] water of monophasic [bmin] [PF_6] tert. butanol system. Both these procedures are applied to substrates using chiral ligands. This procedure permits recycling and reuse of the osmium-liquid catalyst (N. Jain, A. Kumar, S. Chauhan and S.M.S. Chauhan, Tetrahedron, 2005 **61 (5)**, 1015)

Oxidative carbonylation of amines to give phenyl carbamate and diphenyl urea is usually performed by alkali metal containing selenium compounds as catalysts (H.S. Kim, Y.J kim, S.D. Lee and C.S.J. Chin, Catal, 1999, **184**, 526). The main problem in this reaction is the difficulty in separating the catalyst and the product from the reaction mixture. This has been overcome by preparing ionic liquids containing anionic selenium species; these are found to exhibits high activity for the carbonylation of amines at temperature as low as 40° (H.S. Kim, Y.L. Kim, H. Lee, K.Y. Park, C. Lee and C.S. Chin. Angew. Chem. Int. Ed. Engl. 2002, **41**, 4300)

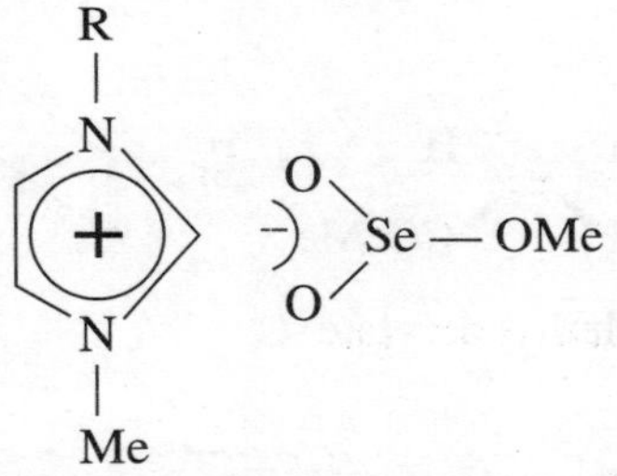

R = Me, Et, Bu

(Selenium-anion based imidazolium ionic liquid)

Wacker-type oxidation of olefins can be performed by $PdCl_2$ immobilised in [bmin] [BF_4] and [bmin] [PF_6] using H_2O_2 as oxidant (R.S. varma, E. Shale-Demessie and U.R. Pillai, Green Chem., 2002, 170).

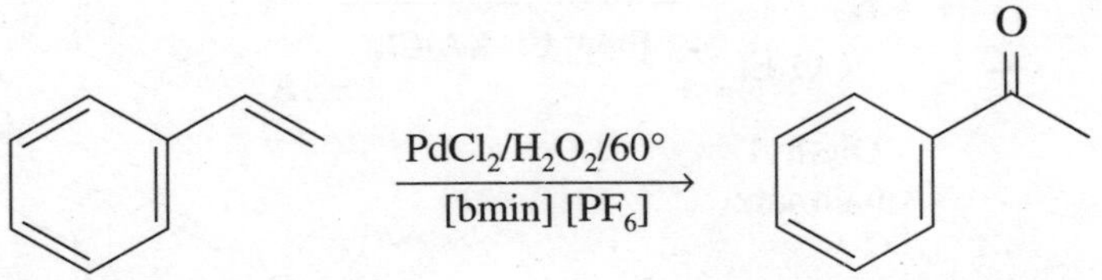

Styrene $\xrightarrow[\text{[bmin] [PF}_6\text{]}]{PdCl_2/H_2O_2/60°}$ Acetophenone

(iv) Fridel Crafts Reaction

The conventional catalyst used in Friedel-crafts alkylation and acylation is $AlCl_3$, which gives rise to disposal problems and formation of byproducts. The use of ionic liquid [e_{min}] Cl-$AlCl_3$ in place of solid $AlCl_3$ enhances the rate of the reaction and selectivity and the ionic liquid also acts as solvent for the reaction (C.J. Adams, M.J. Earle, G. Roberts, K.K. Seddon, Chem. Commun, 1998, 2072;

C.W. Lee, Tetrahedron Lett., 1999, **40**, 2461; A. Stark, B.L Maclean and R.D. Singer, J. Chem. Soc. Dalton Trans, 1999, 63). Friedel-crafts alkylation of aromatic compounds with alkenes using Sc $(OTf)_3$-ionic liquid system gives the advantage of simple procedure, easy recovery and reuse of catalysts. This procedure contributes to the development of environmentally benign and waste free process (C.E. Song, W.H. Shim, E.J. Roh and J.H. Choi, Chem. Commun., 2000, 1695.

(v) Diels-Alder Reaction

Diels-Alder reaction, known for its usefulness is a highly stereospecific reaction giving good yields of the adducts. Ionic liquids like [bmin] $[BF_4]$, [bmin] $[ClO_4]$, [emin] $[CF_3SO_3]$ and $[e_{min}]$ $[PF_6]$ are used for the Diels-Alder reaction. Thus, the reaction of cyclopentadiene and methylacrylate exhibited significant rate enhancements, high yields and strong endo selectivities comparable with the best results obtained in conventional solvents (T. Fischer, T. Sethi, T. Walton and J. Woolf, Tetrahedron Lett., 1990, **40**, 793; M.J. Earle, P.B. Mc Mormac and K.R. Seddon, Green Chem, 1999, **1**, 23)

Cyclopentadiene + Methyl acrylate ($H_2C{=}CH{-}CO_2Me$) $\xrightarrow{IL}$ endo-CO_2Me adduct + exo-CO_2Me adduct

(vi) Knoevenagel Condensation

Ionic liquids act as Lewis acid catalyst and solvent in the Knoevenagel condensation of aromatic aldehydes with diethyl malonate to give benzylidene malonate, which subsequently undergoes **Michael addition** with diethyl malonate.

Aramatic aldehyde (R–C_6H_4–CHO) + Diethyl malonate ($CH_2(CO_2Et)_2$) $\xrightarrow[\text{[bpy] Cl. XAlCl}_3]{\text{[bmin] Cl. XAlCl}_3 \text{ or}}$ R–C_6H_4–CH=C$(CO_2Et)_2$ + R–C_6H_4–CH$(CH(CO_2Et)_2)$–CH$(CO_2Et)_2$ (minor)

(vii) Aldol Condensation

The self condensation of propanal to yield 2-methylpent-2-enal has been carried out in imidazolium ionic liquids (C.P. Mehnert, N.C. Dispenzire, R.A. Cook, Chen. Commun., 2002, 1610)

$$CH_3CH_2CH_2CHO \xrightarrow{[bmin]\ [PF_6]} \left[CH_3CH_2-\underset{}{\overset{OH}{CH}}-\underset{CH_3}{CH}-CHO \right] \xrightarrow{-H_2O} CH_3CH_2CH=\underset{CH_3}{C}-CHO$$

Propanal   2-Methylpent-2-enal

Aromatic aldehydes on reaction with acetone in presence of (S)-proline as catalyst in the ionic liquid [bmin] [PF_6] gave asymmetric aldol product with good enantioselectivity (P. Kotrusz, I. Kmentova, B. Gotov, S. Toma and E. Solcaniova, Chem. Commun., 2002, 2510)

$$R-C_6H_4-CHO + CH_3COCH_3 \xrightarrow[\text{[bmin] [PF}_6\text{]/RT}]{\text{(S)-Proline}} R-C_6H_4-CH(OH)CH_2COCH_3$$

Aromatic aldehydes   Acetone   Aldol product

(vii) Wittig Reaction

It is a convenient procedure for C = C bond formation. In this reaction, use of ionic liquid [bmin] [PF_4] as solvent and using stabilised ylides permits easy separation of the alkenes from Ph_3PO and also recycling of the solvent (V. Le Boulaire and R. Grie, Chen. Commun., 2000, 2195). The alkenes obtained have E-Stereoselectivity, as in the use of other solvents.

(ix) Suzuki Coupling Reaction

This coupling reaction is used for the synthesis of biaryls and is a carbon-carbon bond forming reaction (N. Miyaura and A. Suzuki, Chem, Rev., 1995, **95**, 2457). This reaction using a Pd catalyst can be conveniently accomplished in ionic liquid (C. Zhang, J. Huang, M.L. Trudell and S.P. Nolen, J. Org. Chem., 1999, **64**, 3804)

$$C_6H_5Br + Me-C_6H_4-B(OH)_2 \xrightarrow[\text{[bmin] [BF}_4\text{]}]{(Ph_3P)_4Pd} C_6H_5-C_6H_4-Me$$

Bromobenzene   Toyl boranic acid   p-Methyl biphenyl (92%)

The Suzuki coupling reaction has also been carried out under mild conditions in ionic liquid with methanol as co solvent (necessary to solubilize the phenylboronic acid) using ultrasound (R. Rajgopal, D.V. Jarikote and K.V. Srinavason, Chem. Commun., 2002, 616).

(x) Still Coupling Reaction

It is an important reaction for the preparation of polyarenes and diaryl and aromatic carbonyl compounds (M. Kosugi and K.Fugami, J. Organomet. Chem., 2002, 653, 50). In the normal reaction these is problem of expense of the catalyst. It has been found that the use of palladium complexes immobilized in ionic liquid offer considerable advantage over the classical organic solvents used for still coupling reaction (S.T. Handy and X. Zhang, org. Lett, 2001, **3**, 233)

$$+ \; RSn(Bu)_3 \xrightarrow[\text{[bmin] [BF}_4\text{]}]{\text{Pd(II)CuI/Ph}_3\text{As}}$$

(xi) Henry Reaction

The reaction is an aldol type reaction and consist in reacting nitroalkanes having α-hydrogen atom with carbonyl compounds to give β-hydroxynitro compounds (H. Henry. Compt. Rend. 1895, **120**, 1265) This reaction can be accelerated by chloroaluminate ionic liquids (A. Kumar and S.S. Pawar, J. Mol. Catal. A. 2005, **235** 244)

$R\text{–}CH_2\text{–}NO_2$ + R^1COR^2 $\xrightarrow{IL}$ O_2N, OH, R, R^1, R^2

Nitroalkane Carbonyl compound Nitroalcohols

(xii) Stetter Reaction

The reaction of aldehydes with olefins to give 1,4-dicarbonyls is known as stetter reaction. It is found that the reaction proceeds very well in ionic liquid using Et_3N as catalyst (S. Anjaiah, S. Chandrasekhar and R. Gree, Adv. Synth. Catal. 2004, **346**, 1329)

CHO, F + COOMe $\xrightarrow{IL}$ O, F, F, OH, +

p-Fluoro benzaldehyde Methyl acrylate

IL = [HO–CH₂CH₂-thiazolium(R at C4, N, S)] X

F–C₆H₄–CO–CH₂CH₂–CO–OMe

R = CH_2Ph; X = Cl^-

R = Et, X = Br^-

(xiii) Esterification, Ether Formation and Pinacol-Pinacolone Rearrangement

The ionic liquid TSIL (II) has been successfully used for the above reactions (A. Cole, J.L. Jensen, I. Ntai, KLT. Tran, K.J. Weaver, D.C. Forbes and J.H. Davis, J. Am. Chem. Soc., 2002, **124**, 5962)

$$(\text{Bu})_3\overset{+}{P}(CH_2)_3SO_3H \;\; p\text{-}CH_3C_6H_4SO_3^-$$

TSIL II

Esterification

$$\underset{\text{Caproic acid}}{CH_3(CH_2)_4COOH} + \underset{\text{1-Octanol}}{CH_3(CH_2)_6CH_2OH} \xrightarrow[22°,\ 48\ \text{hr.}]{\text{TSIL II}} \underset{\text{n-Octylcaproate (82\%)}}{CH_3(CH_2)_4CO_2CH_2(CH_2)_6CH_3}$$

Ether Formation

$$\underset{\text{1-Octanol}}{2CH_3(CH_2)_6CH_2OH} \xrightarrow[22\text{-}175°,\ 2\ \text{hr}]{\text{TSIL (II)}} \underset{\text{Dioctyl ether}}{\left[CH_3(CH_2)_6CH_2\right]_2O}$$

Pinacol-Pinacolone Rearrangement

$$\underset{\text{Pinocol}}{Ph_2C(OH)\text{–}C(OH)Ph_2} \xrightarrow[180°,\ 2\ \text{hr}]{\text{TSIL (II)}} \underset{\text{Pinacolone}}{Ph_2C(OH)\text{–}C(=O)Ph}$$

(xiv) Pechmann Condensation

Substituted Coumarians can be synthesised via the Pechmann condensation involving reaction of phenols and ethylacetoacetate by heating in 1 L at 80° for 15 min (Y. Gu, F. Shi and Y. J. Deng, J. Mol. Catal. A. Chem, 2004, **212**, 71).

IL = N(+)N SO_3H $CF_3SO_3^-$

R_1 R_2 OH R_3 + O O CH_3 $\xrightarrow[\text{15 min, 80°}]{\text{IL}}$ R_1 R_2 O O R_3 CH_3

Ethylacetoacetate

R_1	R_2	R_3
H	OH	H
H	OH	OH
H	OMe	H
MeO	OH	H
OH	OH	H
Me	H	H

Conclusion

The properties and behaviour of ionic liquids can be adjusted to suit an individual reaction type. So these are described as designer solvents. A large number of organic reactions can be performed by using ionic liquids. These include C-C bond formation and C-O bond forming reactions. Such reactions are the basis for organic transformations.

Chapter 10

Transformations in Polyethylene Glycol

10.1 ADIPIC ACID

It is obtained by the oxidation of cyclohexene with H_2O_2 in polyethylene glycol (PEG)-Na HSO_4.

$$\text{Cyclohexene} + 4H_2O_2 \xrightarrow[\text{PEG/NaHSO}_4]{\text{Na}_2\text{WO}_4} \text{Adipic acid (COOH, COOH)}$$

Procedure[1]

A mixture of Na_2WO_4. $2H_2O$ (0.186 g, 0.564 m mol), PEG-2000 (12.5 g), $NaHSO_4$ (10 g), hydrogen peroxide (30.0 g 441 m mol) is stirred and cyclohexene (5 g, 60.85 m mol) added. The mixture is stirred at 90° for 8-10 hr. and the solution allowed to stand at 0° for 35 hr. The separated adipic acid is filtered, m.p. 150°

Notes

1. J. Chen, S.K. Spear, J.C. Huddleston, J.D. Holbrey, R.P. Swatloski and R.D. Rogers, Ind. Eng. Chem. Res, 2004, **43**, 5358.
2. Most of the adipic acid is manufactured by the oxidation of cyclohexanol and or cyclohexanone with HNO_3. The emission of N_2O is responsible for environmental pollution. The procedure describe is an elegant, environmentally benin synthesis.

10.2 AZOBENZENE

It is obtained by the oxidation of hydrazobenzene by PEG. NO_2^1.

$$C_6H_5NHNHC_6H_5 \xrightarrow{\text{PEG. NO}_2} C_6H_5N{=}NC_6H_5$$

Hydrazobenzene **Azobenzene**

Procedure[1]

Hydrazobenzene (10 m mol) and PEG. NO_2 (1.9 g, 20 m mol) is stirred at 35°. Cold water (20 mL) is added and the precipitated azobenzene is filtered and crystallised from alcohol, m.p. 67-68°. yield 80%. The filtrate is concentrated under reduced pressure and the residual PEG could be recycled.

Notes

1. R.-Z. Qiou, Y. Zhang, X.-P. Hui, P.-F. Xu, Zi-Yi. Zhang, X.-Y. Wang and Y.-L Wang, Green Chemistry 2001, 3, 186.
2. The required hydrazobenzene can be obtained from nitrobenzene by treatment with magnesium turnings in absolute alcohol in presence of few crystals of iodine (V.K. Ahluwalia, Renu Agarwal, Comprehensive practical organic chemistry (Preparations and quantitative analysis, Universities Press, 2000, Page 34).
3. This is a convenient green route[1] for the synthesis of azocompounds using PEG. NO_2 form hydrazo compounds.
4. The required PEG. NO_2 is obtained[1] by passing dry NO_2 gas (obtained by the reaction of sodium nitrite with dilute H_2SO_4; the formed NO gets converted into NO_2 in presence of air) into polyethylene glycol-(PEG 400). The formed PEG. NO_2 is obtained as an orange liquid.
5. Azo compounds were earlier synthesised by diazo coupling or oxidising hydrazones using NBS, $KMnO_4$ or fuming HNO_3.
6. Using the PEG NO_2, a number of azocompounds can be obtained in good yield.

10.3 TERTIARY BUTYL ALCOHOL

It is prepared[1] by treating tertiary butyl chloride with water in PEG-300.

$$(CH_3)_3CCl + H_2O \xrightarrow[\text{RT. 20 min}]{\text{PEG-300}} (CH_3)_3COH$$

Tert butyl chloride Tert butyl alcohol

Procedure

A mixture is tertiary butyl chloride in water (20 mL) and PEG 300 (5 mL) is stirred at room temperature for 20 min. The reaction mixture is extracted with ether. The ether extract is dried (Na_2SO_4) and distilled to gives tert butyl alcohol (having camphor like odor) b.p. 82-83° in 95% yield (n_4^{20} 0.78581; d_4^{25} 0.78086)

Notes

1. N.F. Leininger, R. Clontz, J.L. Gainer and D.V. Kirwan in Clean solvents, Alternative media for chemical reaction and processing, ed. M.A. Abraham and L. Molns, ACS symposium series 819, American chemical society, Washington DC, 2002, p. 208.
2. Tert Butyl alcohol was earlier obtained by catalytic hydration of isobutylene (U.S. Patent 2,477, 380) (to Atlantic Refining).

10.4 BENZYL ACETATE

It is obtained[1] by treatment of benzyl bromide with sodium acetate in PEG 400.

$$\underset{\text{Benzyl bromide}}{C_6H_5CH_2Br} + \underset{\text{Sod. acetate}}{CH_3COONa} \xrightarrow[\text{RT}]{\text{PEG 400}} \underset{\text{Benzyl acetate}}{C_6H_5CH_2OCOCH_3}$$

This a **substitution reaction** and is performed in PEG 400.

Procedure[1]

A mixture of benzyl bromine (0.05 mol) and anhydrous sodium acetate (0.05 mol) in PEG 400 (5 mL) is stirred for 5 min at room temperature. The reaction mixture is extracted with ether and ether extract washed with water and dried (Na_2SO_4). Distillation of ether gives benzyl acetate, b.p. 213° in 95% yield. The remaining PEG can be used agin for esterification

Notes

1. E. Santaniello, in Crown ethers and phase transfer catalysts in polymer science, ed. L.J. Mathias and C.E. Carraher Jr., Plenum, New York, 1984, P.397.
2. The reaction of benzyl bromide with sodium iodide or sodium cyanide in PEG-400 gives the corresponding iodide (benzyl iodide) and cyanide (benzyl cyanide) in excellent yield. See also preparation benzyl cyanide (Section 5.2)

10.5 BENZYLALCOHOL

It is obtained by the reaction of appropriate polymer-supported succinyl ester followed by heating with ammonium formate resulting in catalytic transfer hydrogenation.

$$\text{PEG 5000} + \underset{\text{Succinic arahydride}}{\begin{matrix} CH_2\text{—}CO \\ | \quad\quad\quad O \\ CH_2\text{—}CO \end{matrix}} \xrightarrow[\Delta\ 24\ \text{hr}]{\text{Disoproyl ethylamine (DIEA), } CHCl_2} \underset{\text{PEG supported succinyl-ester}}{\text{MeO-PEG-O-CO-}CH_2CH_2\text{-COOH}}$$

Procedure[1]

Step (i) Polymer-supported succinyl ester

PEG 500 (2 g) is refluxed with succinic anhydride (4 g, 4.0 m mol) in the presence of disoproylethyl amine (DIEA, 1 mL) in dichloromethane (15 mL) for 24 hr. The solvent is evaporated to one third of its volume and the PEG-Supported ester is precipitated by addition of excess chilled ether (50 mL) followed by filtration and drying in vacuum. Yield 1.92 g (95%).

Step (ii) Polymer supported PEG

Condensation of PEG supported succinyl ester with phenol. Phenol (4.0 m mol), dicyclohexycarbodiamide (DCC, 9.8 m mol) and catalytic amount of dimethylaminopyridine (DMAP, 5 mg) is added to the PEG- supported succinyl ester (1.9 g) in dry dichloromethane (25 mL) and the mixture stirred for 24 hr. under nitrogen. The formed dicyclohexyl urea and excess DCC is removed by filtration. The filtrate is concentrated (valuo) and chilled ether (200 mL) added. The polymer-supported PEG is filtered and dried in valuo, yield 1.8 g (91%). Its ^{1}H-NMR spectra is recorded.

Step (iii) Benzyl alcohol

The polymer-supported substrate (obtained above) is refluxed in methanol with ammonium formate (1 g, 16m mol) and catalytic amount (5% w/w) of Pd/C (10%) for 8 hr. under nitrogen. The Pd/C is filtered, filtrate evaporated and dichloromethane (20 mL) added. The precipitated ammonium formate is filtered, the filtrate concentrated to one third of its original volume. Addition of chilled dry ether (200 mL) yielded cleaved PEG, which is filtered. The filtrate is concentrated and purified by chromatography (Silica gel) 200 mesh, elution with EtOAc/hexane) to afford benzyl alcohol in 85% yield.

Notes

1. M.S. Sampath Kumar, P. Chakrawarthy, S.D. Sawant, P.P. Singh. M.S. Rao and J.S. Yadav, Tetrahedron Lett., 2005, **46**, 3591
2. Though benzyl alcohol can be obtained by a number of procedures but this methadoly demonstrates catalytic transfer hydrogenation.
3. Using the above procedure, $p-C_6H_4CH(OH)-CH_3$, $C_6H_5CH_2CH_2CH_2OH$, $C_6H_5CH=CHCH_2OH$ can also be obtained.

10.6 2-p-CHLOROPHENYL-5-SEMICARBAZIDETHIAZOLIN-4-ONE

It is prepared by the reaction of PEG-supported 1-aminocarbonyl 1,2-diaza-1,3-butadine with phenylthioamide.

PEG supported 1-amiocarbonyl 1,2-diaza-1,3-butadiene + p-chloro phenyl thoamide ($p-ClC_6H_4C(S)NH_2$) → [ ] → ● — OH + 2-p-Chlorophenyl-5-semicarbazide thiazolin-4-one

Procedure[1]

Step (i) PEG-supported-1-aminocarbonyl 1,2-diaza-1,3-butadiene

Poly(ethylene glycol) methyl ether (5 g) (average MW 5000) in toluene (45 mL) is refluxed for 3 hr. in the presence of tert-butylacetoacetate (10 equiv). Diethylyl ether (65 mL) is added to the cooled reaction mixture to give PEG-bound β-ketoester, 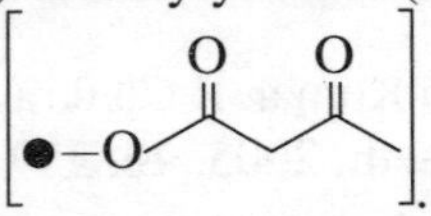 . The separated product is treated with semi carbazide hydrochloride (5 equiv) and sodium carbonate (5 equiv) in methanol (55 mL). The reaction mixture is allowed to stand at room temperature for 5 hr. The PEG-supported hydrazone 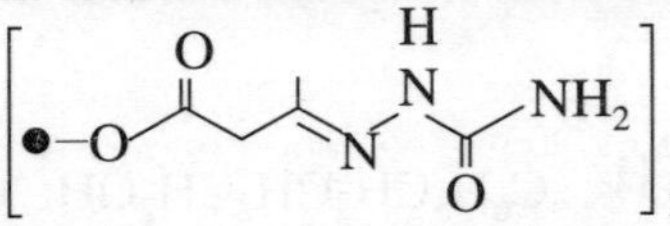 in filtered, and dissolved in methylene chloride (150 mL) and washed with water. The organic layer is dried ($Na_2\ SO_4$) and added slowly with stirring (1.5 hr) to phenyltrimethylammonium tribromide (PTAB) in dichloromethane at room temperature. The product, polymer bound α-bromohydrazone 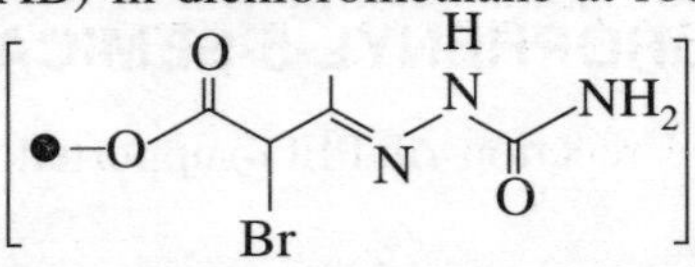 separated is treated with saturated solution of sodium carbonate (2 × 25 mL), organic layer dried ($Na_2\ SO_4$) and solvent evaporated (valuo) and residual oily product macerated with ether (50 mL) to give the required PEG-supportedt-1-amminocarbonyl 1, 2-diaza-1, 3- butadiene. It is washed with ether and used for the next step.

Step (ii) 2-p-Chlorophenyl-5 semicarbazidethiazolin 4-one

A stirred solution of PEG-supported-1-amiocarbonyl 1, 2-diaza-1, 3-butadiene (5 g) (obtained above) in methylene chloride/methanol (8 mL, 1:4) is treated with a solution of p-chorophenylthioamide (3 equiv) in methanol (2 mL). The reaction mixture is stirred for one hr. to give the required 1-p-chorophenyl-5-semicarbazididethiazolin 4-one in 50 % yield (99% purity), m.p. 200-201°. Its IR, ^{1}HNMR and ^{13}C NMR spectra is recorded.

Notes

1. O.A. Altanasi, L.De. Crescentini, G. Faw, P. Fillippone, S. Lilliini, F. Mantellini and S. Santeusanio, Org. Lett., 2005, **7**, 2469.
2. The synthesis of the intermediate, PEG-supported-1-aminocarbonyl-1,2-diazo-1, 3-butadiene is a one pot synthesis and is an environmentally friendly procedure.
3. Using this procedure a larger number of substituted phenyl thiamides can be used to give the corresponding 2-phenyl substituted- thialol-4 ones.

10.7 1,2-DIPHENYL-1,2-ETHANEDIOL

It is obtained by **asymmetric dihydroxylation**[1] of trans stilbene using osmium tetraoxide in PEG in presence of chiral catalyst, $(DHQD)_2$ PHAL.

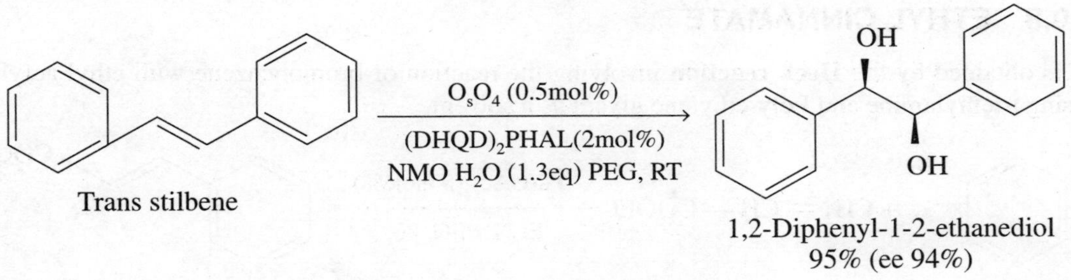

Trans stilbene

O_sO_4 (0.5mol%)
$(DHQD)_2PHAL$(2mol%)
NMO H_2O (1.3eq) PEG, RT

OH
OH

1,2-Diphenyl-1-2-ethanediol
95% (ee 94%)

$(DHQD)_2PHAL$ is represented as given below and is available.

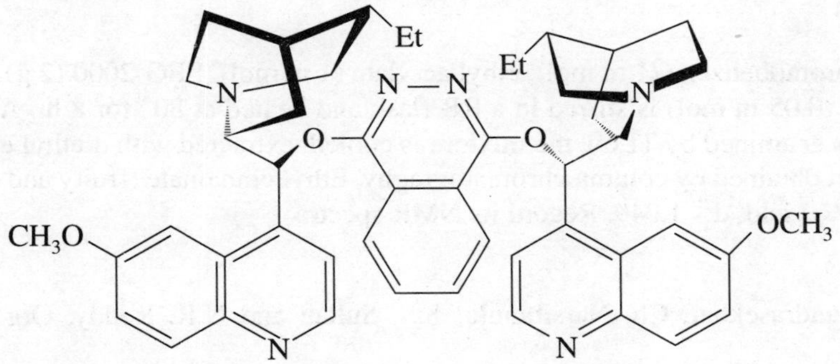

Et
N—N
N
O
Et
N
O
CH_3O
OCH_3
N
N

This reagent is known as **sharpless reagent** and the procedure of dihydroxylation in known as **sharpless procedure**.

Procedure

Trans stilbene (2 m mol) in PEG (2 mL) is added to $(DHQD)_2PHAL$ (2 mol%) and OsO_4 (0.5 mol%). The reaction mixture is stirred for 2 hr, extracted with ether. The ether extract is washed with dilute hydrochloric acid (10%) and then with water and brine. Removal of ether form the extract yield 1, 2-diphenyl-1, 2-ethanediol in 95% yield (94% ee). Record its NMR spectra.

Notes,

1. S. Chandrasekhar, Ch. Narsihmulu, S.S. Sultana and N.R. Reddy, Chem. Commun. 2003, 1716
2. The recovered PEG and the catalyst can be used for other lots of dihydroxylations of olefins
3. Using the above procedure, asymmetric dihydroxylation of ethyl cinnamate and ethyl 4-methoxy cinnamate etc can be effected in good yield and ee.

10.8 ETHYL CINNAMATE

It is obtained by the **Heck reaction** involving the reaction of bromobenzene with ethyl acrylate using triethylamine and Poly-ethylene glycol as a solvent.

$$\text{Bromobenzene} + CH_2{=}CH{-}COOEt \xrightarrow[Et_3N,\ PEG,\ 80^\circ]{Pd(OAc)_2(5\ mole\%)} \text{Ethyl cinnamate}$$

Bromobenzene Ethylacrylate Ethyl cinnamate

Procedure[1]

A mixture of bromobenzene (1 m mol), ethyl acrylate (1 m mol), PEG-2000 (2 g), TEA (1 m mol) and $Pd(OAc)_2$ (0.05 m mol) is stirred in a RB flask and heated at 80° for 8 hr. After the reaction is complete (as examined by TLC), the mixture is cooled, extracted with diethyl etler (3 × 10 mL) and the product obtained by column chromatography. Ethyl cinnamate (fruity and balsamic odour) obtained in 90% yield, d_4^{20} 1.049. Record its NMR spectra

Notes

1. S. Chandrasekhar, Ch. Narsihmulu, S.S. Sultan and N.R. Reddy, Org. Lett., 2002, **4**, 4399.
2. Ethyl cinnamate is also obtained by esterification of cinnamic acid (with ethanol), which in turn is obtained by **Perkin reaction** from benzaldehyde acetic anhydride and potassium acetate (Org. Reactions, 1942, **I**, 248).
3. Using styrene or n-butyl vinyl ether in the above Heck reaction, the product obtained are stilbene (10 hr heating) or 1-(2-butoxy)-(E)-1-ethenyl) benzene (12 hr heating) in 93 and 88% yield respectively.
4. Using p-methoxy benzaldehyde or p-chlorobenzaldehyde and heating with ethyl acrylate, styrene and n-butyl vinyl ether, the corresponding products are obtained in 80-95% yield.
5. PEG not only acted as an efficient solvent medium but also as a PTC for smooth C-C-bond formation.

$$R{-}C_6H_4{-}Br + CH_2{=}CH{-}X \xrightarrow[Et_3N,\ PEG,\ 80^\circ]{Pd(OAc)_2} R{-}C_6H_4{-}CH{=}CH{-}X$$

R = Cl, OMe X = Ph, CO_2Et, nBuO and/or R (branched product with X)

6. The PEG can be recycled in the same Heck reaction

10.9 ETHYL 3-HYDROXY-3-PHENYL-2-METHYLENE PROPANATE

It is prepared by the reaction of benzadehyde with ethyl acylate in presence of DABCO using PEG as a solvent at room temperature.

$$C_6H_5CHO + CH_2{=}CHCO_2Et \xrightarrow[\text{PEG, RT}]{\text{DABCO (20 mol\%)}} C_6H_5CH(OH)C({=}CH_2)CO_2Et$$

Benzaldehyde — Ethyl acrylate — Ethyl 3-hydroxy-3- phenyl-2-methylene propanate

The reaction is known as **Baylis-Hillmann reaction.** It uses PEG 400 as a rapid and recyclable reaction medium using the conventional basic catalyst DABCO with vey good yield. In this case recyclability is achieved with no further addition of DABCO to the reaction medium. In fact, DABO is recycled for the first time in this transformation[1].

Procedure

A mixture of benzaldehyde (0.01 mol) and ethyl acrylate (0.01 mol) in PEG-400 is stirred is presence for DABCO (20 mol%) at room temperature for 2 hr. The product is isolated by ether extraction (5 × 10 mL). The ether extract is evaporated and the crude product purified by column chromatography in 92% yield.

Notes

1. S. Chandrasekhar, Ch. Narsihmulu, B. Saritha and S.S. Sultana, Tetrahedron Lett., 2004, 45, 5865.
2. In this reaction besides ethylacrylate a number of other olefins like acrylonitrile, methyl vinyl ketone etc can be used. Even different aldehydes like p-nitro, p-fluorobenzaldehydes can be used.
3. See also 3-hydroxy-3-phenyl-2-methylene propanamide (Section 2.23)

10.10 (4 S)-HYDROXY-4-(4′-NITROPHENYL)-BUTAN-2-ONE

It is obtained by **asymmetric transfer aldol reaction**[1] between p-nitrobenzaldehyde and acetone in presence of L-proline (as a catalyst) using Polyethylene glycol (PEG-400) as solvent.

$$O_2N{-}C_6H_4{-}CHO \xrightarrow[\text{acetone (4 eq.) PEG, r.t. 30 min}]{\text{L-proline (10 mol\%)}} O_2N{-}C_6H_4{-}CH(OH)CH_2COCH_3$$

p-Nitrobenzadehyde — (4S)-Hydroxy-4-(4′-nitrophenyl)-butane-2-one (94%)

Procedure[1]

To a stirred solution of L-Proline (10 mol %) in PEG (2 mL) is added acetone (4 m mol) at room temperature (inert atmosphere). The mixture is stirred for 5 min, p-nitrobenzaldehyde (1 mmol) added, and mixture stirred for 30 min. The reaction mixture is extracted with ether, solvent removed and the product purified by silica gel column chromatography to give pure (4 S)-hydroxy-4-(4′-nitrophenyl)-butan-2-one (in 94% yield as pale brown viscous oil.

1. S. Chandrasekhar, N.R. Reddy, S.S. Sultana, Ch. Narsihmulu and K. V. Reddy, Tetrahedron 2006, **62**, 338
2. The reaction can also be performed by using diacetone alcohol as a source of acetone
3. This is an important C-C bond forming reaction
4. L-Proline is an efficient catalyst for the direct asymmetric aldol reaction of aromatic and aliphatic aldehydes in PEG and can be recycled along with PEG.

10.11 3(S)-HYDROXY-3-PHENYL PROPANONE-2

It is obtained by L-Proline catalysed direct **asymmetric aldol reaction** of benzaldehyde and acetone in PEG-400.

Benzaldehyde —[L-proline (10 mol%) Acetone (4 eq.); PEG, R.T. 30 min]→ 3(S)-Hydroxy-3-phenyl propanone-2

Procedure[1]

Acetone (4 m mol) is added to a stirred solution of L-proline (10 mol%) in PEG (2 mL) at room temperature under inert atmosphere. After stirring for 5 min, benzaldehyde (1 m mol) is added, mixture stirred for 30 min and anhydrous ether (5 mL) added. The stirring is continued for 5 min. The ether layer separated and the mother liquior (PEG + proline) extracted again with ether (2 × 5 mL). Ether is removed (vacuo) and the 3(S)-hydroxy-3-phenyl propanone-2 obtained is purified is column chromatography (silica gel). yield 58%, 58% ee

Notes

1. S. Chandrasekhar, N.R. Reddy, S.S. Sultana, Ch. Narsihnulu and K.V. Reddy, Tetrahedron, 2006, **62**, 338
2. The above asymmetric aldol reaction could also be conducting by using diacetone alcohol in place of acetone giving good results
3. Much better yields (80-90%) can be obtained by using o-, m and p-nirobenzaldehydes, 2-chlorobenzaldehyde

4. Normally aldol condensation reaction is known to be effective in C-C bond forming process and is commonly used in organic synthesis. Usually aldehydes containing α-hydrogen atom undergo aldol condensation to give β-hydroxyaldehydes called aldol.

$$CH_3CHO + CH_3CHO \xrightarrow{NaOH} CH_3CH(OH) - CH_2 - CHO$$

5. Asyrnmetric direct asymmetric aldol reaction of acetone with various aromatic and aliphatic aldehydes in PEG-400 gives asymmetric aldol products

10.12 4-METHYLBYPHENYL

It is obtained by the reaction of p-methyl phenylboronic acid and bromobenzene in PEG-400 in presence of $PdCl_2$, KF by irradiation in a MW oven for 50 see.

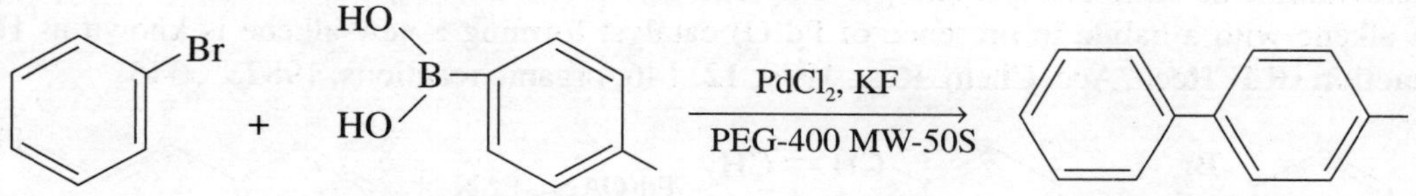

Br HO HO B + $\xrightarrow[\text{PEG-400 MW-50S}]{PdCl_2,\ KF}$

Bromobenzene p-Methyl phenyl boronic acid 4-Methyl biphenyl 80%

The reaction is known as microwave accelerated **Suzuki Cross-coupling reaction**

Procedure[1]

A mixture of PEG 400 (2 g), palladium chloride (0. 050 g, 0.282 m mol), p-methylphenylboronic acid (0.15 g, 1.10 m mol) and bromobenzene (0.156 g, 1 m mol) and potassium fluoride (0.39 g) is heated in an unmodified house hold microwave oven at 240 W for 50 sec. The reaction mixture is cooled, extracted with ether (3 × 3 mL). The extract is chromatographed to give 4-methylbiphenyl, m.p 44-45° in 84% yield. Its NMR spectra is recorded.

Notes

1. V.V. Namboodiri and R.S. Varma, Green chemistry, 2001, 3, 146
2. Biphenyl was also prepared by sonication of bromobenzene in THF in presence of lithium (Section 6.2)
3. The above procedure is an environmentally friendly method. In this procedure the catalyst and PEG can be recycled in other similar reaction.
4. Using the procedure a number of biphenyls can be prepared in 70-90% yield. Some examples include

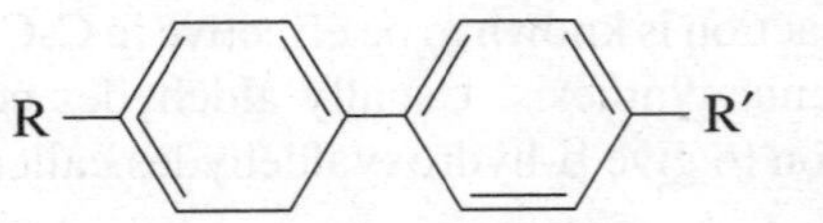

R	R′
CH_3	CH_3
CH_3	CHO
CH_3	OMe
CH_3	COMe
CH_3	NO_2
CH_3	F

5. In place of phenylboronic acid, olter substituted boronic acids can also used be to give the corresponding biphenyls in 75-85% yield.

10.13 STILBENE

It is obtained by the reaction of bromobenzene with styrene in presence of Pd $(OAc)_2$ using triethylamine as basic catalyst and polyethylene glycol (PEG-2000) as solvent. This coupling of an alkene with a halide in presence of Pd(O) catalyst forming a new alkene is known as **Heck reaction** (R.F. Heck, Acc. Chem. Res., 1979, **12**, 146; organic reactions, 1982, , 345)

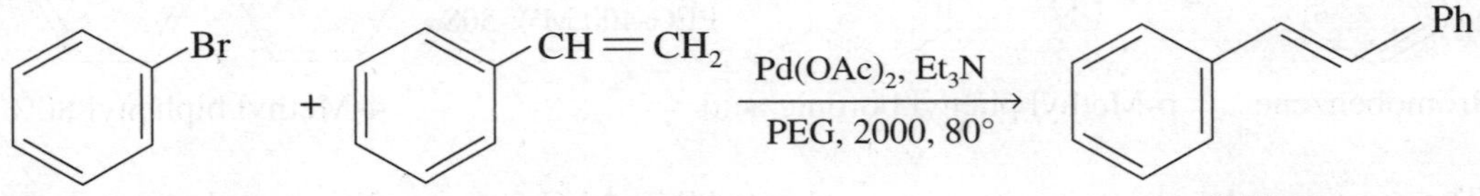

Br + CH=CH_2 $\xrightarrow[\text{PEG, 2000, 80°}]{Pd(OAc)_2,\ Et_3N}$ Ph

Bromo benzene Styrene Stilbene

Procedure[1]

A mixture of bromobenzene (1 m mol) styrene (1 m mol), PEG-2000 (2 g), TEA (1 m mol) and Pd $(OAc)_2$ (0.05 m mol) is stirred and heated at 80° for 10-12 min. The reaction mixture is cooled, extracted with ether (3 × 10 mL) and purified by column chromatography. Yield 93% M.P. 124-125°

Notes

1. S. Chandrasekhar, Ch. Narsihmulu, S.S. Sultana and N.R. Redy, Org. Lett, 2002, **4**, 4399.
2. In the above reaction PEG acted as an efficient solvent and also as a PTC for smooth C-C-bond formation
3. In place of bromobenzene (in the above reaction) p-methoxy bromobenzene, 3,4-methylenedioxybromobenzene and p-chlorobromobenzene gave the corresponding styrene's in 80-90% yield
4. Use of ethyl acrylate in the above reaction gave ethyl cinnamate in 90% yield.
5. Normally stilbene is obtained by the reaction of triphenyl phosphonium salt with benzaldehyde (V.K. Ahluwalia and Renu Aggarwa, Comprehensive Practical organic chemistry, Preparation quantitative analysis, Universities press, 2000, Page 219

$$C_6H_5CH_2Cl + (C_6H_5)_3P \xrightarrow{\bar{O}C_2H_5} (C_6H_5)_3P{=}CH{-}C_6H_5$$

$$\xrightarrow[\text{warm}]{C_6H_5CHO} (C_6H_5)(H)C{=}C(H)(C_6H_5) + (C_6H_5)_3P{=}O$$

6. Heck reaction can also be preformed in ionic liquid as the solvent (W.A. Herrmann and V.P.W. Bohm, J. Organometal. Chem. 1999, **572**, 141) (See Section 9.5)

10.14 STYRENE DIOL

Osmium tetraoxide in catalytic amount is commonly used for the **dihydroxylation of alkenes.** However, the main problem in the dihydroxylation is the contamination of the product with osmium besides its cost and toxicity. This restricts the use of osmium tetraoxide in industry for the manufacture of pharmaceuticals and fine chemicals.

An efficient dihydroxylation of alkenes uses catalytic amount of OsO_4 (0.5 mol%) in PEG-400 as solvent and N-methylmorpholine (NMO) as the reoxidatiant[1]. Using this procedure styrene is converted in styrene diol.

$$\text{Styrene} \xrightarrow[\text{NMO } H_2O \text{ (1.3 eq)} \\ \text{PEG-400, RT}]{O_sO_4\text{(0.5 mol\%)}} \text{Styrene diol}$$

Procedure[1]

Osmium tetraoxide (2.5 mg, 0.5 mol%) is added to a stirred solution of styrene (208 mg, 2 m mol) in PEG 400 (2 g) and NMO. H_2O (351 mg, 2.6 m mol) under inert atmosphere. The mixture is stirred for 2 min. and extracted (by decantation) with ether (5 × 5 mL). The total ether extract is washed with water, brine and dried (Na_2SO_4). The ether is removed (rotary evaporator) and the residual styrene diol is chromatographed (column chromatography) in 94% yield. Record its NMR spectra. The PEG layer (left after ether extraction) is reused for subsequent dihydroxylation.

Notes

1. S. Chandrasekhar, Ch. Narsihmulu, S.S. Sultana and N.R. Reddy, Chem. Commun., 2003, 1716
2. In this procedure the formed diol is free of any contamination with osmium:
3. Using the above procedure, a number of alkenes like α-methyl styrene, stilbene, ethyl cinnamate, and cyclohexene could be dihydroxylated is 90-95% yield.

$$R\text{-}CH{=}CH\text{-}R' \xrightarrow[\text{NMO.H}_2\text{O(1.3eq)}\ \text{PEG. RT.}]{\text{OsO}_4\text{(0.5 mol\%)}} R\text{-}CH(OH)\text{-}CH(OH)\text{-}R'$$

R = alkyl, aryl
R′ = H, alkyl, aryl

4. See also asymetric dihydroxylation (Section 10.7)

10.15 MISCELLANEOUS APPLICATION OF PEG IN ORGANIC TRANSFORMATION

(i) Diel-Alder Reaction

The reaction of 2, 3-dimethyl -1, 3-butadiene with acrolien in PEG 300 gave the Diels-Alder adduct in 95% yield.

$$CH_2{=}C(CH_3)\text{-}C(CH_3){=}CH_2 + CH_2{=}CHCHO \xrightarrow{\text{PEG 300}} \text{Adduct}$$

2, 3-Dimethyl 1,3-butadiene   Acrolein   Adducl 95%

The yield is 14 times in comparison to the reaction carried out in methanol (N.F. Leininger, R. Clontz, J.L. Gainer and D.V. Kirwan in clean solvents. Alternative media for chemical reactions and processing, ed. M.A. Abraham and L. Moens, ACS symposium series 819, American chemical Soc., Washington DC. 2002, P.208)

The Diels. Alder reaction of 2, 3-dimethyl 1, 3-butadiene with nitrosobenzene is PEG 300 exhibited more than 3 times increase in yield compared to the same reaction in dichloromethane.

(ii) Sandmeyers Reaction

The reaction normally involves decomposition of diazonium salts in presence of CuCl. However, the yields are poor.

$$Ar\overset{+}{N_2}\overset{-}{Cl} \xrightarrow{\text{CuCl}} ArCl + N_2$$

The reaction of diazonium salt with an halide like NaCl in PEG gives the corrsponding chloride in excellent yields. (N. Suzuki, T. Azuma, Y. Kaneko, Y. Izawa, H. Tamioka and N. Nomoto, J. Chem. Soc. Perkin Trans, I, 1987, 645)

$$ArN_2Cl + NaCl \xrightarrow[\text{R.T., Stirring}]{\text{PEG}} ArCl$$

Using this procedure a number of aryl chlorides can be obtained. Use of NaCN in place of NaCl gives the corresponding cyanides in excellent yield.

(iii) Heck Coupling

PEG is a very useful reaction medium for heck coupling reaction which is a Pd-catalysed C-C-bond forming reaction. (S. Chandrasekhar, Ch. Narsihmulu, S.S. Sultana and N.R. Reddy, Org. Lett., 2002, 4399). For more details see ethyl cinnamate preparation (Section 10.8)

(iv) Baylis-Hillman Reaction

The reaction involving condensation of aromatic aldehodes with ethyl acrylate, acrylonitrite or methyl vinyl ketone to give new C-C-bond formation has been carried with in PEG as a solvent in presence of DABCO (S. Chandrasekhar, Ch. Narsihmulu, B. Saritha and S.S. Sultana, Tetrahedron Lett., 2004, **45**, 5865). For more details see preparation of ethyl 3-hydroxy-3-phenyl-2-methylene propanate (Section 10.9)

(v) Suzuki Cross Coupling Reaction

It involves the synthesis of biaryls by the reaction of aromatic aldehydes with organoboranes is presence of base. This is a carbon-carbon bond forming reaction and can be conveniently performed in PEG (V.V. Nanboodiri and R.S. Varma, Green Chemistry, 2001, **3**, 146). For more details sec preparation of 4-methyl biphenyl (Section 10.12)

(vi) Asymmetric Aldol Condensation

The reaction of aldehydes with acetone takes place in presence of L-proline in PEG at room temperature to yield Chiral β-hydroxyketones (S. Chandrasekhar, N.R. Reddy, S.S. Sultana, Ch. Narsihmulu and K.V. Reddy, Tetraheron, 2006, **62**, 338).

$$\underset{\text{Aromatic aldehyde}}{\text{Ar — CHO}} \xrightarrow[\text{acetone (4 eq), PEG, R.T. 30 min}]{\text{L-Proline (10 mol\%)}} \underset{\text{Chiral β-hydroxy ketone}}{\text{Ar–CH(OH)–CH}_2\text{–CO–CH}_3}$$

For more details, see preparation of 3(S)-hydroxy-3-phenylpropanone-2 (Section 10.11) and (4S)-hydroxy-4-(4′nitrophenyl)-butan-2-one (Section 10.10)

(vii) Oxidations

A number of oxidation reactions can be conducted in PEG-200 and PEG-400. Some examples are given below:

(a) Oxidation of benzyl bromide

(E. Santaniello, A. Manzocchi and P. Sozzani, Tetrahedron Lett., 1979, **20**, 4581)

$$\underset{\text{Benzyl bromide}}{C_6H_5CH_2Br} \xrightarrow{K_2Cr_2O_7/\text{PEG-400}} \underset{\text{Benzaldehyde (90\%)}}{C_6H_5CHO}$$

(b) Dihydroxylation of olefins

(S. Chandrasekhar, Ch. Narsihmulu, S.S. Sultana and N.R. Reddy, Chem. Commun., 2003, 1716)

$$\underset{\text{Stilbene}}{C_6H_5CH{=}CH\,C_6H_5} \xrightarrow[\text{OsO}_4\text{, PEG-400, 2-4 hr}]{\text{N-methyl morpholine (NMO)}} \underset{\text{Stilbene diol}}{C_6H_5\underset{|}{\overset{}{C}}H(OH) - CH(OH) - C_6H_5}$$

See also styrene diol (Section 10.14)

(viii) Reductions

PEG has been used in a number of reduction. Some examples are give below.

(a) Reduction of carbonyl compounds

(E. Santaniello, A. Manzocchi and P. Sozzani, Tetrahedron Lett. 1979, **20**, 4581)

$$CH_3COC_6H_{13} + NaBH_4 \xrightarrow{\text{PEG-400}} CH_3CHOHC_6H_{13}$$

(b) Reduction of esters

(E. Santaniello, P. Ferraboschi and P. Sozzani, J. Org. Chem., 1981, **46**, 4584)

$$R - COOR' + NaBH_4 \xrightarrow{\text{PEG-400}} R - CH_2OH$$

R = alkyl, aryl
$R' = CH_3, C_2H_5$

(c) Reduction of halides of the type R-CHX-R′

(E. Santaniello, A. Fiecchi, A. Manzocchi and P. Ferraboschi, J. Org. Chem., 1983, **48**, 3074)

$$R - CHXR' + NaBH_4 \xrightarrow{\text{PEG-400}} R - CH_2 - R'$$

R = alkyl, aryl
X = Cl, Br, I
$R' = H, CH_7C_4H_9$

(d) Reduction of acyl chlorides

(E. Santaniello, A. Fiecchi, A. Manzocchi and P. Ferraboschi, J. Org. Chem., 1983, **48**, 3074)

$$R - COCl + NaBH_4 \xrightarrow{\text{PEG-400}} R - CHO$$

$R = C_6H_5$, p-BrC_6H_4

(e) Partial reduction of alkynes to cis-olefins

$$\underset{\text{Alkynes}}{R-C\equiv C-R'} \xrightarrow[\text{PEG-400}]{\text{Pd}-\text{CaCO}_3/\text{PbO (Linders catalyst)}} \underset{\text{Cis olefins}}{R-CH=CH-R'}$$

(ix) Williamson ether synthesis

PEG has the ability to serve as PTC because the polyethylene oxide chain can form complexes with metal cations as in the case of crown ethers. A typical application is the Willianson ether synthesis involving reaction of an alkyl halide and an alcohol

$$R-OH + R'X \xrightarrow{\text{PEG}} R-O-R'$$

Another example is

$$C_{10}H_{21}\ OH + \underset{X = \text{Cl, Br, I}}{C_4H_9X} \xrightarrow{\text{PEG 300-2000}} C_{10}H_{21}-O-C_4H_9$$

(B. Abribat, Y.Le. Bigot and A. Gaset, Synth. Commun., 1994, **24**, 2091; Tetrahedron, 1996, **52**, 8245)

10.16 CONCLUSION

Polyethylene glycol and its solutions are believed to be green reaction medium of future. Its special features are low toxicity, low volatility and biodegradability and its low cost. PEG has been used as a reaction medium for a large number of organic transformations. It has been used for a number of carbon-carbon bond forming reaction like Heck reaction, Baylis-Hillmann reaction, asymmetric transfer aldol reaction, asymmetric aldol reaction, Suzuki cross-coupling reaction, dihydroxylation of alkenes, Diels-Alder reaction, Sandmeyers reaction, oxidations and reductions. PEG also finds use as a PTC in reaction like Williamson ether synthesis.

Chapter 11

Transformations Involving Polymer Support Reagents or Substrates

In this chapter, preparation of some regeants using polymer support is described. These regeants in turn can be used for various transformations.

11.1 BROMINATION OF CUMENE USING POLY-N-BROMO-SUCCINIMEDE (PNBS)

PNBS is a novel and efficient polymer based brominating agent. It is used as benzylisc and allylic brominating agent. Thus bromination of cumene gives[1] α, β, β′-tribomocumene, compared to bromination with NBS, which give α-bromo and α, β-dibramocumene

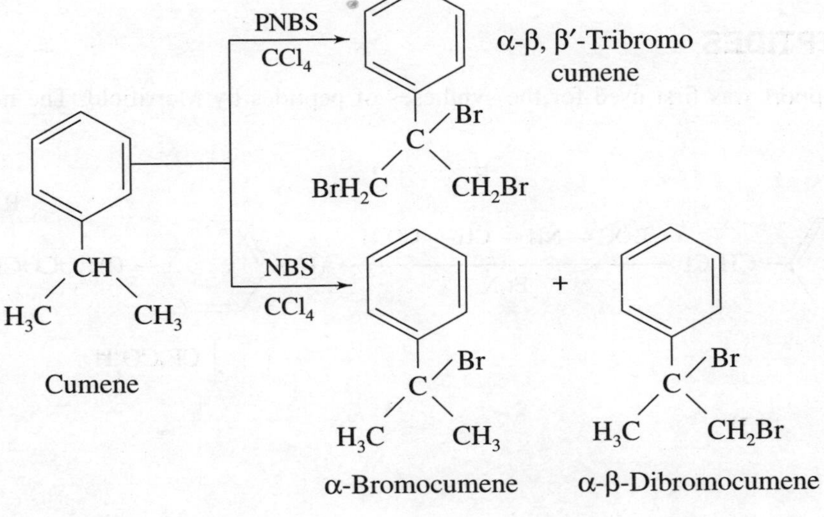

PNBS, CCl_4

α-β, β′-Tribromo cumene

Br

C

BrH_2C CH_2Br

CH

H_3C CH_3

Cumene

NBS, CCl_4

\+

Br

C

H_3C CH_3

α-Bromocumene

Br

C

H_3C CH_2Br

α-β-Dibromocumene

Notes

1. C. Yaroslavsky, E. Patchornik and E. Katchalski, Tetrahedron Lett., 1970, 3629; Isrel J. Chem., 1970, 37.
2. The polymer PNBS is obtained by the addition of bromine to a suspension of polymalemide polymer in aqueous NaOH solution. The polymalemide is obtained by free radical polymerisation of meleimide in the presence of 2.5-5% divinylbenzene.

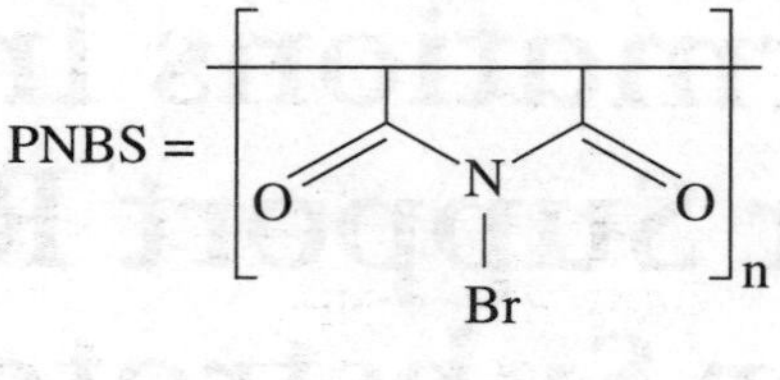

PNBS = [maleimide ring: O, N, O; N–Br]$_n$

11.2 BROMOPOLYSTYRENE

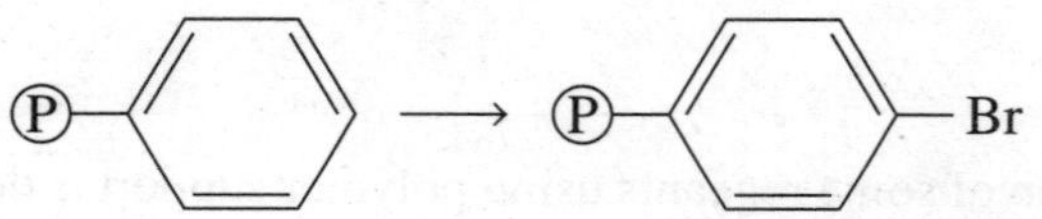

Ⓟ–C$_6$H$_5$ ⟶ Ⓟ–C$_6$H$_4$–Br

Bromine (13.6 g) in carbon tetrachloride (20 mL) is added slowly to polystyrene cross-linked with 1% DVB (20 g) and thallic acetate [Tl $(OAc)_3$]. The mixture is stirred for 1 hr. at RT and then refluxed for 1.5 hr. The polymer is finally washed with dil. HCl-dioxane, water, ethanol, dioxane and ether and dried to get bromopolystyrene

Note

1. M.J. Farall, J.M.J. Frechet, J. Org. Chem, 1976, **41**, 3877; F. Camps, J. Castells, M.J. Ferrando and J. Font, Tetrahedron Lett, 1971, 1713.

11.3 PEPTIDES

Polymer support was first used for the syntheses of peptides by Merrifield. The main steps are given below:

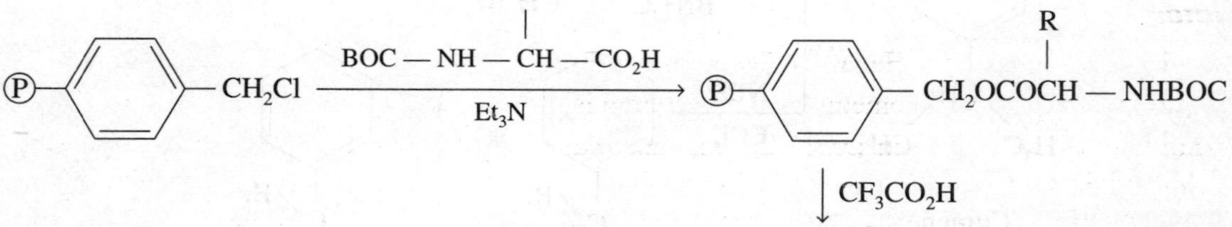

Ⓟ–C$_6$H$_4$–CH_2Cl $\xrightarrow[Et_3N]{BOC—NH—CH(R)—CO_2H}$ Ⓟ–C$_6$H$_4$–$CH_2OCOCH(R)$—NHBOC

↓ CF_3CO_2H

$$\text{Ⓟ}-C_6H_4-CH_2OCOCH(R)-NH-CO-CH(R')-NHBOC \xleftarrow[\text{DCC}]{BOC-NH-CH(R_1)-CO_2H} \text{Ⓟ}-C_6H_4-CH_2COCH(R)-\overset{+}{N}H_3$$

$$\downarrow CF_3CO_2H$$

$$\text{Ⓟ}-C_6H_4-CH_2OCOCH(R)-NHCOCH(R')-\overset{+}{N}H_3 \xrightarrow{\text{HF}} \text{Ⓟ}-C_6H_4-CH_2F + H_2N-CH(R')-CONH-CH(R)-COOH$$

11.4 POLYSTYRENE ACETIC ACID

It is obtained by the hydrolysis of cyanomethylated polystyrene.

$$\text{Ⓟ}-C_6H_4-CH_2CN \longrightarrow \text{Ⓟ}-C_6H_4-CH_2COOH$$

The cycanomethylated polystyrene (4.85 g) is suspended in a mixture of conc. H_2SO_4 (15 mL), acetic acid (15 mL) and water (15 mL). The mixture is heated with stirring at 115-120° for 8-10 hr. The resin is filtered, washed with water (5 × 10 mL), dioxane-water (3:1, 3 × 10 mL), dioxane (3 × 12 mL), dioxane-ethanol (1:3, 3 × 12 mL), ethanol (3 × 12 mL) and ether (3 × 10 mL). It is finally dried in vacuo over P_2O_5, yield 5.0 g.

Notes

1. T. Kusama and H.Hayatsu, Chem. Pharm. Bull, 1970, 18, 319.
2. The required cyanomethylated polystyreni is prepared as follows.

Soldium cyanide (2 g) in DMF (20 mL) and water (4 mL) is added to a stirred suspension of Merrifieldresin (a 2% cross linked copolymer of polystyrene and divinyl benzene and is commercially available). The stirred mixture is heated at 115-120° for about 20 hr. The resin is filtered, washed with water, dioxane-water (1: 3), dioxane, dioxane-ethanol (2:1), ethanol and ether. The resin is finally dried in vaceo over P_2O_5.

11.5 POLYSTYRENE-ALUMINIUM CHLORIDE

It is prepared as follows (D.C. Neckers, D.A. Koistra and G.W. Green, J. Am. Chem. Soc., 1972, 9284)

$$\text{Ⓟ}-C_6H_5 \xrightarrow{AlCl_3} \text{Ⓟ}-C_6H_4-AlCl_3$$

11.6 POLYSTYRENE BORONIC ACID

It is obtained[1] by the reaction of trimethyl borane with lithigated polystyrene.

$$\text{Ⓟ}-C_6H_4-Li + (CH_3)_3B \longrightarrow \text{Ⓟ}-C_6H_4-B(OH)_2$$

Trimethyl borane (18 mL) is added to a suspension of lithigated polystyrene. The mixture is stirred (24 hr), liquid phase removed and the resin is washed with THF. The resulting mixture is heated with stirring for 1.5 hr. with dioxane (140 mL), water (12 mL) and HCl (36 mL). The resin is filtered, washed with dioxane-water (3 : 1), dioxane, acetone and methanol. yield 12.96 g

Notes

1. M.J. Farrall, J.M.J. Frechet, J. org. Chem, 1976, **41**, 3877
2. The required lithiated polystyrene is obtained as follows:

Brominated polystyrene (Section 11.2) (15 g) is suspended in cyclohexane (TME DA) (4 ml). The mixt is stirred and n-Buli (13.5 ml, 2.5 m) is added. The mixture is heated for 4.5 hr. at 65°. The liquid phase is removed and the resin is washed with cyclohexane to yield lithiated resin.

11.7 POLYSTYRENE PERACID

It is obatined[1] form polystyrene carboxylic acid by reacting with H_2O_2 is presence of p-toluenesulfonic acid.

$$\text{Ⓟ}-C_6H_4-COOH \xrightarrow[PTS]{H_2O_2} \text{Ⓟ}-C_6H_4-CO_2OH$$

It can also be obtained by treating polystyrene acid chloride with sodium peroxide and H_2O_2

Notes

1. J.M.J. Frechet and K.E. Haque, macromolecules 1975, 8, 130.

11.8 WITTING REACTION[1]

It is carried out as shown below.

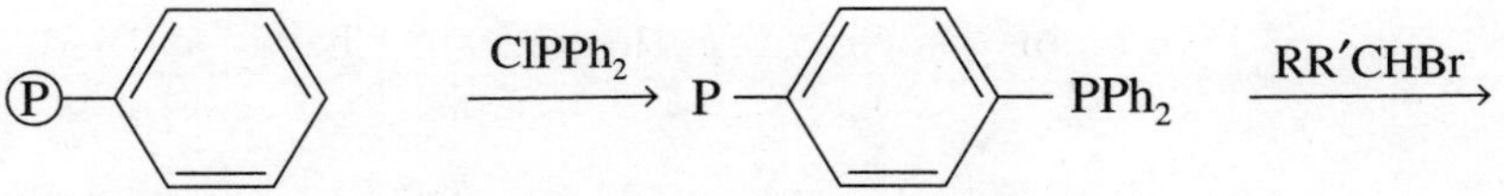

$$\text{Ⓟ}-C_6H_5 \xrightarrow{ClPPh_2} \text{P}-C_6H_4-PPh_2 \xrightarrow{RR'CHBr}$$

Lithiated Polystyrene[1] Polystyrene phosphine

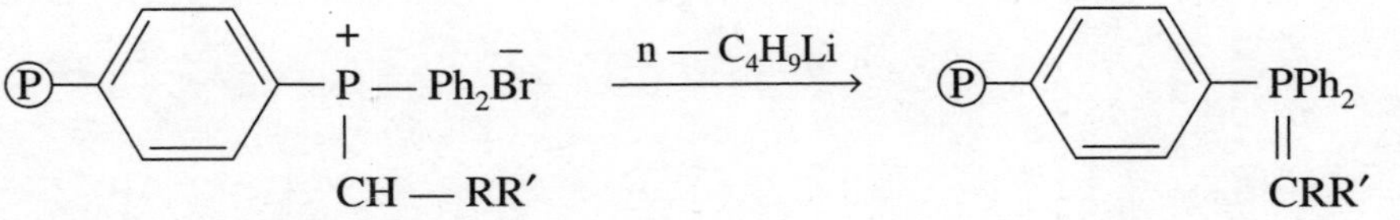

$$\text{Ⓟ}-C_6H_4-\overset{+}{P}(CH-RR')-Ph_2\overset{-}{Br} \xrightarrow{n-C_4H_9Li} \text{Ⓟ}-C_6H_4-PPh_2{=}CRR'$$

Polymeric phosphorium bromide Polystyrene Wittig reagent

The polystyrene witting reagent is reacted with other carbonyl compounds like benzaldehyde, p-chlorobenzaldehyde to give the alkenes.

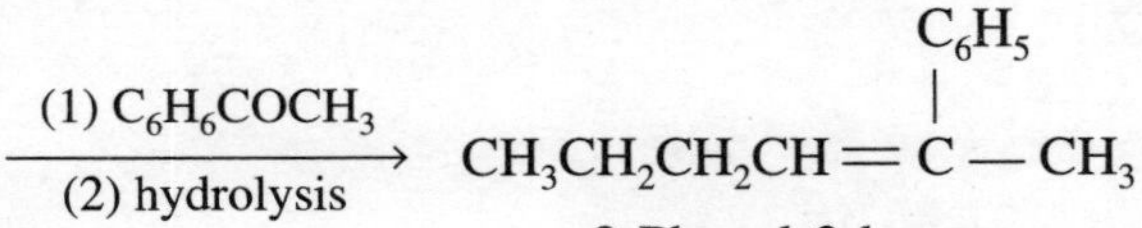

$$\xrightarrow[\text{(2) hydrolysis}]{\text{(1) } C_6H_6COCH_3} CH_3CH_2CH_2CH{=}\underset{}{\overset{C_6H_5}{\overset{|}{C}}}-CH_3$$

2-Phenyl-2-hexene

Notes

1. W. Heitz and R. Michels, Angew. Chem. Int. Ed., 1972, **11**, 298; S. V. Mckiney and J.W. Rakshys, J. Chem. Soc. Chem. Commun, 1972, 134; F. Camps, J. Castells, J. Font and F. Vela, Tetrahedron lett, 1971, 1715.

11.9 CONCLUSION

Use of polymeric materials in organic synthesis facilitates the isolation and purification in simple way and in high yields. The polymeric reagents are useful for a number of organic transformations.

Chapter 12

Miscellaneous Transformations

12.1 2-ACETYLCYCLOHEXANONE

It is obtained by the reaction of cyclohexanone with pyrrolidine or morpholene in presence of p-toluenesulphonic acid or triethylamine. The formed enamine is reacted with acetic anhydride or acetyl chloride to yield 2-acetylcyclohexanone

Cyclohexanone + Pyrrolidine $\xrightarrow{\text{p-Toluene sulfonic acid}}$ Enamine $+H_2O$

CH_3—C_6H_4—SO_3H

Enamine $\xrightarrow{\text{base}}$ … + $CH_3-C(=O)-Cl$ (Acetyl chloride) $\xrightarrow{-Cl^-}$

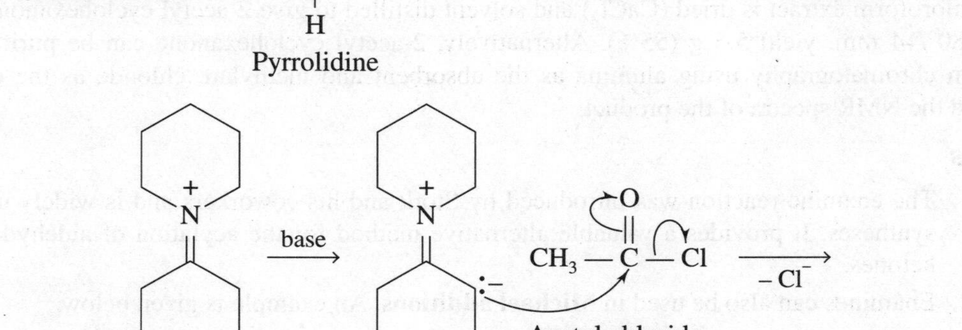

$\longrightarrow$ [iminium intermediate: $\overset{+}{N}$ pyrrolidinium ring on cyclohexane bearing $C(=O)-CH_3$] $\xrightarrow{H_2O}$ [cyclohexanone with $C(=O)-CH_3$]

2-Acetyl cyclohexanone

Materials

Cyclohexanone	15.5 mL
Pyrrolidine	16.8 mL
Triethylamine	10 mL
Acetyl chloride	18 mL
Chloroform dry	125 mL

Procedure

The examine is first obtained by refluxing a solution of pure cydohexanone (15.5 mL, 0.15 mol) and pyrrolidine (16.8 mL 0.23 mol) is dry benzene (100 mL) using a Dean-stark apparatus. When the reaction is complete (TLC), the solvent is removed by distillation and the examine (12-14 g) obtained is used as such for the next step.

A mixture of examine (12 g, 0.07 mol), triethylamine (10 mL) and dry chloroform (80 mL) is shaken to obtain a homogeneous solution (anhydrous conditions). A solution of acetyl chloride (18 mL, 0.25 mol) in dry chloroform (45 mL) is added with constant shaking to the enamine solution. The mixture is refluxed for 2 hrs (water-bath). Water (20 mL) and conc. HCl (20 mL) is added to the reaction mixture and refluxing continued for 3 hr. more. The chloroform layer is separated, washed with water (2 × 50 mL). To total aqueous solution is neutralised with sodium bicarbonate solution keeping the mixture slightly acidic. It is extracted with chloroform (2 × 25 mL). The chloroform extract is dried ($CaCl_2$) and solvent distilled to give 2-acetyl cyclohexanone, b.p. 160-180°/14 mm, yield 5.5 g (55%). Alternatively, 2-acetyl cyclohexanone can be purified by column chromatography using alumina as the absorbent and methylare chlonde as the eluant. Record the NMR spectra of the product.

Notes

1. The enamine reaction was introduced by Stork and his coworkers and is widely used in syntheses. It provides a valuable alternative method for the acylation of aldehydes and ketones.
2. Enamines can also be used in **Michael additions**. An example is given below.

$$\text{(pyrrolidine enamine of cyclohexanone)} + CH_2{=}CHCN \xrightarrow[\text{reflux}]{C_2H_5OH} \text{(iminium)}\ CH_2CH_2CN \xrightarrow{H_2O}$$

$$\longrightarrow \text{2-}(CH_2CH_2CN)\text{cyclohexanone}$$

3. Using the above procedure, 2-ethyl and 2-allyl cyclohexanones can also be prepared.

12.2 ALKYL AZIDES

These are conveniently prepared[1] by sonication of a solution at alkyl halide (e.g., benzyl bromides 2.1 m mol) and sodium azide (2.76 m mol) in water containing pillared clay for 2.5-4.5 hr. at 0-5°. The reaction was worked up by adding water, extraction with ether and subsequent distillation

$$\text{(R}_1\text{,R}_2\text{,R}_3\text{,R}_4\text{-benzyl bromide)} + NaN_3 \xrightarrow[\text{Pillared clay,))))\ 2.5-4.5 hr., 0°}]{H_2O} \text{benzyl-}N_3$$

R_1	R_2	R_3	R_4
H	H	H	H
CH_3	H	H	H
H	H	CH_3	H
H	H	H	CH_3
H	H	H	NO_2

Notes

1. R.S. Varma, K.P. Naicker and Dalip Kumar, J. Molecular catalysis. A: Chemical, 1999, **149**, 153.
2. The reaction can also be conduced by stirring at 90-100° without sonication
3. Using this method, alkyl and allyl azides can also be prepared (H. Prtebe, Acta. Chem. Scand. ser. B. 1984, **38**, 895.

12.3 3-ARYL-2H-1,4-BENZOXAZINES

The benzoxazines, known for there anti-inflammatory activity were prepared in low yields (D.R. Sridhar, C.V. Reddy Sastry, O.P. Bansal and R. Pulla Rao, Ind J. Chem., 1983, **22B**, 297; 1981, 11, 912; P. Battistoni P. Bram and G. Fava, Synthesis, 1979, 220). There are conveniently obtained in good yield by the condensation of appropriately substituted 2-aminophenol with phenacyl bromide in presena of a PTC (tetrabutyl ammonium hydrogen salphate) in the presence of aqueous potassium carbonate [Y. Sabitha and A.V. Subba Rao, Synth. Communication, 1987, **17(3)**, 341]

$$\text{R-}C_6H_3(OH)(NH_2) + C_6H_5COCH_2Br \xrightarrow[\text{aq. } K_2CO_3]{\text{PTC}} \text{3-Aryl-2H-1,4-benzoxazine } (C_6H_5)$$

Phenacyl bromide

3-Aryl-2H-1, 4-benzoxazines (60%)

12.4 BARBITURIC ACID

It is obtained by heating diethyl malonate with urea in presence of conc. H_2SO_4

$$O{=}C(NH_2)_2 + CH_2(COOC_2H_5)_2 \xrightarrow[\Delta]{H_2SO_4} \text{Barbituric acid} \rightleftharpoons \text{2,4,6-trihydroxypyrimidine}$$

Urea

Barbituric acid

Materials

Sodium	2.87 g
Absolute alcohol	120 mL
Diethyl malonate	20 g (19 mL)
Urea	7.5 g

Procedure

Sodium (cut into small pieces, 2.87 g, 0.125 mol) is added to absolute alcohol (60 mL) in R.B. flask filted with a reflux condenser and a $CaCl_2$ guard tube. After all the sodium has reacted, diethyl malonate, (19 mL, 0.125 mol) and dry urea (7.5 g) in absolute alcohol (60 mL) is added. The mixture is refluxed at 110° (oil bath) for 6 hr. To the reaction mixture hot water (125 mL, 50°) and conc. HCl (25 mL) is added (the resultant solution must be acidic). The mixture is cooled, separated barbituric acid filtered, washed with cold water, M.P. 245° (decomp) yield 12.5 g (78%)

12.5 BENZIDINE

Benzidine (4,4′-diaminobiphenyl) is obtained by the treatment of hydrazobenzene with hydrochloric acid and stannous chloride. This acid catalysed rearrangement of hydrazobenzene to 4,4[1]-diaminobiphenyl is known as **Benzidine rearrangement**[1]

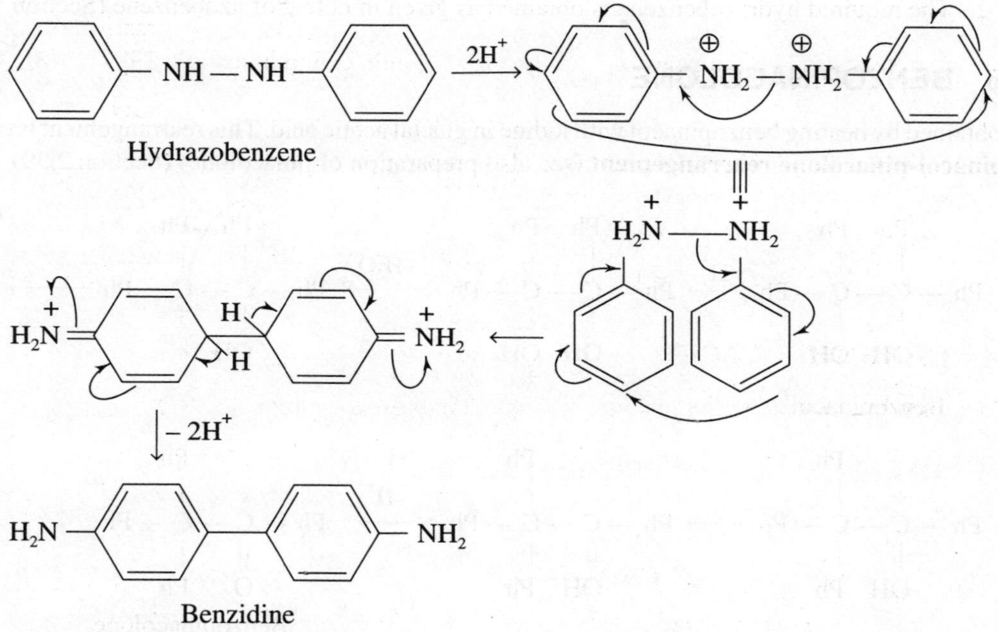

Hydrazobenzene

Benzidine

Materials

Hydrazobenzene	10 g
Methanol	75 mL
Conc. HCl	25 mL
Stannous chloride solution (obtained by dissolving 3 g stannous chloride dihydrate in 3 mL conc. HCl and 10 mL water)	3 mL

Procedure

A mixture of methanol (75 mL), water (40 mL) conc. HCl (25 mL) and stannous chloride solution (3 mL) is cooled (0-2°, ice bath) and then poured on to hydrazobenzene (10 g) in a RB flask. The mixture is stirred (15 min.) in an ice bath and then gently heated to dissolve the benzidine dihydrochlorde. The reaction mixture is cooled (ice-salt mixture) and separated benzidine dihyclrochlorcle is filtered, washed once with dil. HCl and then with methanol. Yield 14 g. It is dissolved in water (100 mL) (if necessary by heating) and sodium bicarbonate solution (7.2 g $NaHCO_3$ in 60 mL H_2O) is added dropwise till the solution is alkaline to litmus. The mixture is cooled (ice-water), separated benzidine filtered, washed with water and recrystallesid from dilute alcohol and dried (vacuum dessicator), yield 5.5 g (55%), m.p. 128°.

Notes

1. T. Sheradsky and S. Auromork's-Grisarv, J. Heterocyclic Chemistry, 1980, 17, 189.
2. The required hydrazobenzene is obtained as given in note 2 of azobenzene (Section 10.2)

12.6 BENZOPINACOLONE

It is obtained by heating benzopinacol with iodine in glacial acetic acid. This rearrangement is called the **pinacol-pinacolone rearrangement** (sec also preparation of pinacolone) (Section 2.39)

$$\underset{\text{Benzopinacol}}{\mathrm{Ph{-}C(Ph)(OH){-}C(Ph)(OH){-}Ph}} \xrightarrow{H^+} \mathrm{Ph{-}C(Ph)(OH){-}C(Ph)(\overset{+}{O}H_2){-}Ph} \xrightarrow{-H_2O} \mathrm{Ph{-}C(Ph)(OH){-}\overset{+}{C}(Ph){-}Ph} \longrightarrow$$

$$\mathrm{Ph{-}\overset{+}{C}(OH){-}C(Ph)_2{-}Ph} \longrightarrow \mathrm{Ph{-}C({=}\overset{+}{O}H){-}C(Ph)_2{-}Ph} \xrightarrow{-H^+} \underset{\text{Benzopinacolone}}{\mathrm{Ph{-}C({=}O){-}C(Ph)_2{-}Ph}}$$

Materials

Benzopinacol	2.5 g
Iodine solution (0.015 M in glacial acetic acid)	12.5 mL

Procedure

A solution of benzopinacol (2.5 g) and iodine solution (12.5 mL) in glacial acetic acid is refluxed for 5 min. The solution is cooled, separated benzopinacolone filtered, washed with cold glacial acetic acid (2 mL). Record the yield, M.P. 182°.

Notes

1. Since it is a rearramgement reaction, so there is 100% atom economy and so it is a green reaction

12.7 BUT-2-EN-1-OL (CROTYL ALCOHOL)

It is obtained by **Meerwein-Ponndorf-Verley reduction**[1] of croton aldehyde with aluminium isopropoxide.

$R{=}CH_3CH{=}CH$

Croton aldehyde ($RCHO$) + Aluminium isopropoxide ($[(CH_3)_2HCO]_2Al{-}O{-}CH(CH_3)_2$) $\rightleftharpoons$ $[(CH_3)_2HCO]_2Al{-}O{-}CH_2R$ + $H_3C{-}CO{-}CH_3$

$$\left[(CH_3)_3CHO\right]_2 AlOCH_2R \xrightarrow{R-\overset{O}{\overset{\|}{C}}-H} \left[RHCHO\right]_3 Al + 2(CH_3)_2CO$$

$$\left[RHCHO\right]_3 Al \xrightarrow{\overset{+}{H_3O}} 3RCH_2OH + Al^{3+}$$

Materials

Coroton aldehyde	21 g
Aluminum isopropoxide	Prepared from aluminum foil (4.7 g) and isopropyl alcohol (50 mL)

Procedure

Prepare aluminium isopropoxode [by refluxing clean aluminum foil (4.7 g), dry isopropyl alceohol (50 mL) and mercuric chloride (0.5 g) in a R.B. Hask (250 mL capacity) under anhydrous conditions. When the liquid is boiling, carbon tetrachloride (2 mL, a catalyst for reaction between Al and alcohols) is added. The heating is continued till all aluminum has reacted. The reaction is finally refluxed for 5 hrs and the remaining solution is used as such Dry isopropyl alcohol (100 mL) and redistilled crotonaldehyde (b.p. 102-103°, 21 g) is added and the mixture is heated at 110° (oil bath). The formed acetone is allowed to distil away. When no more acetone distilled (as indicated by negative DNPH test), the remaining isopropyl alcohol is removed under reduced pressure. To the cooled reaction mixture is added sulphuric acid (3 M, 150 mL form 24.2 mL conc. H_2SO_4 and 133 mL water). The upper oily lager is separated (separatory funnel) and distilled at 100°/120 mm to get crotyl alcohol, yield 11 g (50%), b.p. 119-121°.

Notes

1. M. Meerwein, R. Schmidt, Ann., 1925, **444**, 221, W. Ponndorf, Angew. Ctiem., 1926, **39**, 138; A. Verlay, Bull. Soc. Chim. France, 1925, **37**, 537, 871).

2. Aluminum isopropoxide is a mild reducing agent for reducing carbonyl compounds in good yield and is especially valuable since other groups eg, a conjugated double bond, a nitro or halogen are uneffected.

12.8 o-CHLOROBENZALDEHYLE

It is obtained by the oxidation of o-chlorotoluene with chromyl chloride. The reaction is known as **Etard Reaction**

$$\text{o-Chlorotoluene } (C_6H_4(CH_3)Cl) \xrightarrow{CrO_2Cl_2} \text{o-Chlorobenzaldehyde } (C_6H_4(CHO)Cl)$$

Procedure

o-Chlorotoluene (9.2 mL, 0.079 mole) is dissolved in carbon disulphide (50 mL) in a R B flask (250 mL capacity). Chromyl chloride (25 g) is added to the reaction mixture. The mixture is allowed to stand in cold (using external cooling by cold water) for 3 days. The separated solid is filtered and heated with water (100 mL). The separated o-chlorotoluene (as an oil) on distillation gave pure produced, yield 5.5 g (50%). BP. 209-215°.

Notes

1. A.L. Etard, Canptd. Rend., 1880, **90**, 534; A.L. Etard, Ann. Chem. Phys, 1881, **22**, 218

12.9 CINNAMALDEHYDE

It is obtained by oxidation of cinnamyl alcohol with chromium trioxide-pyridine complex. This oxidation is known as **Sarett oxidation**[1].

$$\underset{\text{Cinnamyl alcohol}}{C_6H_5CH{=}CH\,CH_2OH} \xrightarrow{CrO_3\text{-pyridine complex}} \underset{\text{Cinnamaldehyde}}{C_6H_5CH{=}CH{-}CHO}$$

Materials

Chromium trioxide	1 g
Pyridine	20 mL
Cr O_3-pyridine complex (prepared from 1 g CrO_3)	
Cinnamyl alcohol	1 g

Procedure

A solution of cinnamyl alcohol(1 g) in pyridine (10 mL) is mixed with CrO_3-pyridine complex [obtained by adding chromium trioxide (1 g) in portions to pyridine (10 mL) at 15-20° with stirring]. The reaction flask is stoppered and contents thoroughly mixed. The reaction mixture is allowed to stand overnight, poured into water and extracted with ether. The ether extract is dried ($MgSO_4$), concentrated and distilled. Cinnamaldehyde is collected at 248°, yield 0.68 g (70%).

Notes

1. G. I. Poss, G. E. Arth, A. E. Beyler and L. H. Sarett, J. Am. Chem. Sos., 1953, **75**, 422; J. R. Holum, J. Org. Chem., 1961, **26**, 4814)
2. The oxidation of cinnamyl alcohol to cinnamyal aldehyde can also be effected by **Jones oxidation** involving oxidation with CrO_3 and conc. H_2SO_4 (V.K. Ahluwalia and R. Aggarwal, Comprehensive Practical Organic Chemistry, Universities Press, 2004, Page 137)

12.10 CINNAMIC ACID

It is obtained by heating benzaldehyde with malonic and in presence of ethanoloc ammonia solution. The reaction is known as **knoevenagel condensation**[1]. When this reaction takes place in presence it pyridine or a base, decarboxylation usually occurs. This is known as **Doebner modification**

$$\underset{\text{Cinnamaldehyde}}{C_6H_5CHO} + \underset{\text{Malonic acid}}{CH_2(CO_2H)_2} \xrightarrow[\Delta]{\text{base}} \underset{\text{Cinnamic acid}}{C_6H_5CH{=}CH{-}COOH}$$

Malonic acid $\xrightarrow{\text{base}}$ $H{-}\overset{\ominus}{C}(CO_2H)_2$, which attacks the carbonyl carbon of $R{-}C(=O){-}H$:

$$R{-}CH(\ddot{O}^{\ominus}){-}CH(CO_2H)_2 \xrightarrow{H^+} R{-}CH(OH){-}CH(CO_2H)_2 \xrightarrow[\Delta]{-CO_2,\ -H_2O} \text{Cinnamic acid}$$

Materials

Benzaldehyde	5.3 g
Malonic acid	5.4 g
Ethanolic ammonia solution (8%)	25 mL

Procedure

A mixture of benzaldehyde (5.3 g, 0.005 mol), malonic acid (5.4 g, 0.005 mol) and ethanolic ammonia solution (8%, 25 mL) is heated (water-bath) to get a clear solution. Excess alcohol is distilled off and the residue is heated at 150° until CO_2 ceases to evolve. The residual product is dissolved in water, solution acidified (HCl), cooled and the separated cinnamic acid filtered. It is washed with water and crystallised form hot water, M.P. 134°, yield 5.5 g (75%).

Notes

1. K. Bowder, M. Heilbron, E.R.H. Jones and B.C.L. Weodon, J. Chem. Soc., 1946, 39; K. Knoevenagel, Ber., 1898, **31**, 2596.
2. Cinnamic acid can also be obtained by **Perkins Reaction** involving heating benzaldehyde with acetic anhydride in presence of fused sodium acetate (W. H. Perkins, J. Chem. Soc. 1868, **21**, 53, 181; 1877, **31**, 388)
3. Doebner modification is also useful for the preparation of 3, 4-methylenedioxycinnamic acid, m.p. 238°

Piperonal + $CH_2(COOH)_2$ $\xrightarrow[\text{Piperidine, 95-100°}]{\text{Pyidine}}$ 3,4-Methylenedioxycinnamic acid (H_2C (O)(O) ring, CH=CHCOOH)

Piperonal — Malonic acid — 3,4-Methylenedioxycinnamic acid

12.11 DIMEDONE

It is obtained by **Michael condensation** involving reaction of mestiyl oxide with diethyl malonate in presence of sodium methoxide.

$$CH_2(CO_2Et)_2 + \overset{\ominus}{O}Et \rightleftharpoons :\overset{\ominus}{C}H(CO_2Et)_2 + EtOH$$

$$(CH_3)_2C{=}CH{-}\overset{O}{\overset{\|}{C}}{-}CH_3 \;+ :\overset{\ominus}{C}H(CO_2Et)_2 \rightleftharpoons$$

$$(CH_3)_2\underset{CH(CO_2Et)_2}{\underset{|}{C}}{-}CH{=}\overset{:\ddot{O}:^{\ominus}}{\overset{|}{C}}{-}CH_3 \xrightleftharpoons{C_2H_5OH} (CH_3)_2\underset{CH(CO_2Et)_2}{\underset{|}{C}}{-}CH{=}\overset{O-H}{\overset{|}{C}}{-}CH_3$$

$$\downarrow -H^+ \;\; \bar{O}Et$$

$$(CH_3)_2C(EtO_2CCH)-CH_2-C{=}O,\ CH_2-C({=}O)OEt \xrightarrow[\text{Claisen condensation}]{\overset{\ominus}{O}Et \text{ internal}} (CH_3)_2C(CH(CO_2Et)_2)-CH_2-C({=}O)-CH_3$$

$$\downarrow -\overset{\ominus}{O}Et$$

$$(CH_3)_2C(EtO_2C\,CH)-CH_2-C{=}O,\ CH_2-C{=}O \xrightarrow[(2)\ H^+]{(1)\ KOH} \text{Dimedone}$$

Dimedone

Materials

Sodium	2.3 g
Absolute alcohol	40 mL
Mesityl oxide	10 g (11.6 mL)
Diethyl malonate	17 g (16.1 mL)

Procedure

Clean sodium (2.3 g, 0.1 mol), cut into small pieces is added to absolute alcohol (40 mL) contained in a R.B. flask (250 mL capacity) fitted with a reflux condenser (anhydrouss conditions). When all the sodium has reacted, diethyl malonate (16.1 mL, 0.1 mol) is added (using a dropping funnel). Subsequently, mesityl oxide (11.6 mL, 0.1 mol) is added. The reaction mixture is stirred for 2 hrs under refluxing. Potassium hydroxide solution (12.5 g in 60 mL water) is added, and the mixture refluxed for 2 hrs. more (water bath). The mixture is acidified with dil HCl (55 mL, 1 : 1). Alcohol is distilled (water bath) and the residual solution is acidified again with dilute HCl until acidic to methyl orange. The separated dimedone is filtered, wash with water, dried and crystallised from acetone-petroleum ether, m.p. 147-1400°. Yield 12 g (86%).

12.12 2,3-DIPHENYLQUINOXALINE

It is prepared by heating o-diaminobenzene with benzil

$$C_6H_4(NH_2)_2 + O{=}C(C_6H_5)-C(C_6H_5){=}O \longrightarrow \text{2,3-Diphenylquinoxaline}$$

o-Diaminobenzene   Benzil   2,3-Diphenylquinoxaline

Materials

o-Diaminobenzene	1.65 g
Benzil	3.15 g
Rectified spirit	12 mL

Procedure

A solution of o-diaminobenzene (1.650, 0.015 mol) in ractified spirit (12 mL) is added to a warm solution of benzil (3.15 g, 0.015 mol) in rectified spirit (12 mL). The mixture is warmed for 30 min. (water bath). Water is added dropwise to the above solution unitil slight cloudiness persists. The solution is cooled, separated 2,3-diphenylquinoxaline filtered and crystallised form alcohol, m.p. 125-126°. yield 2.75 g (51%).

Notes

1. The reaction could easily be conducted in a microwave oven by heating for 40 seconds.

12.13 ESTERIFICATION

Normally esterification of a carboxylic acid (RCOOH) with an alcohol (R^1OH) to gives the ester ($RCOOR^1$) is carried out in presence of conc. H_2SO_4, HCl gas or p-toluene sulphonic acid. Esterification is also achived in much better yield by azcotropic distillation of the formed water. Sometimes esters are also obtained by alcoholsis of cynanides ($RCH_2CN + R^1OH + H_2SO_4 \rightarrow RCH_2COOR^1 + NH_4HSO_4$). (For details sec V.K. Ahluwalia and Renu Aggarwal, Comprehenseve Practical Organic Chemistry: Preparation and Quantutative analysis, Universitis press, 2000 P. 84-88.)

Esterification of carboxylic acids and alcohols in presence of conc. H_2SO_4 (catalytic amount) has conveniently been effected by heating is a sealed tube in a micro wave oven for 6 min. (R.N. Gedye, F.E. Smith and K.C. Westaway, can. J. Chem., 1988, 66, 17; R.N. Gedye, W. Rank and K.C. Westaway, Can. J. Chem., 1991, **69**, 706)

$$\text{RCOOH} + \text{R'OH} \xrightarrow[\text{H}_2\text{SO}_4,\ 6\text{ min}]{\text{MW}} \underset{70\text{-}80\%}{\text{RCOOR'}}$$

$$\text{R} = \text{Ph} \qquad \text{R'} = \text{CH}_3,\ \text{CH}_3\text{CH}_2,\ \text{CH}_3\text{CH}_2\text{CH}_2$$

The esterification can also be effected by the sonication of a mixture of carboxylic acid and alcohol in presence of conc. H_2SO_4 (J.M. Khurana, P.K. Sahoo and G.C. Maikap, Synth. Commun., 1990, 2267)

$$\text{RCOOH} + \text{R'OH} \xrightarrow[))))]{\text{H}_2\text{SO}_4,\ \text{RT}} \underset{70\text{-}80\%}{\text{RCOOR'}}$$

A converient method of esterification is the reaction of carboxylic and with alkyl halides, in the presence of triethyl amine (H.E. Hennis, L.R. Thompson and J.P. Long, Ind. Eng. Chem. Prod. Res. Dev. 1968, **7**, 96; H.E. Hennis, J.P. tsterly, L.R. Collins, 1968, **7**, 96; H.E. Hennis, and L.R. Thompson, Ind. Eng. Chem. Prod. Res. bev., 1967, **6**, 193). In this case the PTC is obtained in situ by the reaction of triethylamine and alkyl halide.

$$\underset{}{Et_3N} + \underset{\text{Alkyl halide}}{RX} \longrightarrow \underset{\text{generated PTC (in situ)}}{Et_3\overset{+}{N} + R\overset{-}{X}}$$

$$\underset{\text{Carboxylic acid as sod-salt (aq. soln.)}}{RCOONa} + \underset{\text{Alkyl halide}}{R'X} \xrightarrow{PTC} \underset{\text{ester}}{RCO_2R'} + NaX$$

In place of triethyl amine, quaternary ammonium or phosphonium salt can be directly used in the estrification of carboxylic and with alkyl halides (R. Holmberg and S. Hansen, Tetrahedral lett., 1975, 2307.

Crown ethers have also been used for esterification. Thus, p-bromophenacyl esters can be prepared in 92% yield by the reaction of p-bromophenacyl bromide with potassium salt of a carboxylic and using 18-crown-6 as the solubilizing agent (H.D. Durst, Tetrahedron lett., 1974, 2421)

$$\underset{\text{p-Bromophenacyl bromide}}{p\text{-}BrC_6H_4\text{-}C(=O)\text{-}CH_2Br} + RCOO_2^- \xrightarrow[CH_3CN]{K^+Crown} \underset{\text{p-Bromophenacyl esters (92\%)}}{p\text{-}BrC_6H_4\text{-}C(=O)\text{-}CH_2COOR}$$

Esterification also proceeds very well by heating carboxylic acid and alcohol in an ionic liquid at 22° for 48 hr (A. Cole, J.L. Jensen, I. Ntai, K.L.T. Tran, K.J. Weaver, D.C. Forbes and J.H. Davis, J. Am. chem Soc., 2002, **124**, 5962)

$$I.L = \;\gt\overset{+}{P}(CH_2)_3SO_3H \;\; p—CH_3C_6H_4SO_3^-$$

n-BUTYLACETATE

It is prepared by the reaction of acetic acid with n-butyl alcohol in presence of conc. H_2SO_4.

$$CH_3-\overset{O}{\overset{\|}{C}}-OH \overset{H^+}{\rightleftharpoons} \left[CH_3-\overset{\overset{+}{OH}}{\overset{\|}{C}}-\ddot{O}H \longleftrightarrow R-\overset{OH}{\overset{|}{\underset{+}{C}}}-OH\right] \underset{\text{n-Butyl alcohol}}{\overset{CH_3CH_2CH_2CH_2OH}{\rightleftharpoons}}$$

Acetic acid

$$CH_3-\overset{OH}{\overset{|}{\underset{\underset{Bu\quad H}{O^+}}{\underset{|}{C}}}}-OH \rightleftharpoons CH_3-\overset{\overset{H\ \ +}{O-H}}{\overset{|}{\underset{OBu}{\underset{|}{C}}}}-OH \overset{-H_2O}{\rightleftharpoons} \left[CH_3-\overset{+}{\underset{OBu}{\underset{|}{C}}}-OH \updownarrow CH_3-\underset{OBu}{\underset{|}{C}}=\overset{+}{O}H\right]$$

$$CH_3-\overset{O}{\overset{\|}{C}}-OCH_2CH_2CH_2CH_3 \overset{-H^+}{\rightleftharpoons}$$

Materials

Acetic acid (glacial)	16 mL
n-Butyl alcohol	23.5 mL
Conc. H_2SO_4	0.25 mL.

Procedure

Conc. H_2SO_4 (0.25 mL) is added slowly to a solution of glacial acetic acid (16 mL, 0.26 mol) and n-butyl alcohol (23.50 mL, 0.26 mol) contained in a R.B flask (100 mL capacity). The mixture is refluxed for 6 hr. and pound into water (75 mL). It is extracted with ether, extract washed with sodium bicarbonate solution (20 mL) and water (20 mL). It is dried (Na_2SO_4) and distilled to give n-butyl acetate, b.b. 124-25° yield 11.2 g (80%).

Notes

1. The esterification could be performed by heating in a microwave over for 2-3 min.
2. Using sonication, the reaction is completes in 4-5 min.
3. n-Butyl acetate can also be prepared by heating methyl cyanide with n-butyl alcohol in presence of conc. H_2SO_4

$$CH_3C\equiv N + nBuOH \longrightarrow \left[CH_3-\underset{OBu}{\underset{|}{C}}=\overset{+}{N}H\right] \xrightarrow{H_2O} CH_3-\underset{OC_4H_9}{\underset{|}{C}}=O$$

4. Etherification of acods by refluxing with alcohol in the presence of hydrogen chloride gas is known as **Fisher-speier esterification** (E. Fischer and A. Speier. Ber., 1895, **28**, 3252)
5. In place of conc. sculpuric acid, esterification can also be performed in presence of p-toluene sulphonic acid as the catalyst.
6. All the above notes (1 to 4) hold good for all esterification (given below).

Ethylbenzoate

It is obtained by refluxing benzoic and (15 g. 0.123 mol), absolute ethyl alcohol (32 mL) in presence of conc. H_2SO_4 (0.5 mL) (4 hr refluxing). Working up the reaction mixture as to the case of n-butyl acetate gives ethylbenzoale, b.p. 213° in 80% yield.

$$C_6H_5COOH + C_2H_5OH \xrightarrow{H^+} C_6H_5COOC_2H_5$$

Benzoic acid Ethyl alcohol — Ethylbenzate

Methyl Salicylate

Methyl salicylate, also known as **oil of wintergreen** is a familiar-smelling ester isolated from wintergreen plant (goul theria) in 1843. It is found to have analgesic and antipyretic character as in case of salicylic acid. In fact, the medicinal character of methyl salicylate is due to the case with which it is hydrolysed to salicylic acid under alkaline conditions in the intestinal tract. Salicylic acid is known to have analgesic and antipyretic properties. The ester beside having medicinal properties, is also used to a small extent as a flavoring principle due to its pleasant odour.

Methyl salicylate is prepared by refluxing a solution of salicylic and (17 g, 0.123 mol), absolute methyl alcohol (30 mL) and conc. H_2SO_4 (4 mL, added dropwise with caution) for 4 hrs under anhydrous conditions. The reaction mixture is worked up as in the case of n-butyl acetate. Methyl salicylate, b.p. 223-25° is obtained in 86% yield.

$$C_6H_4(OH)COOH + CH_3OH \xrightarrow{H_2SO_4} C_6H_4(OH)COOCH_3$$

Salicylic acid — Methyl salicylate

Sec. Butyl acetate

It is obtained by passing dry hydrogen chlorde gas into dry butan-2-ol (23 mL, 0.25 mol) until 0.7 g is absorbed. To this solution is added glacial acetic acid (30 mL) and the solution refluxed for 10 hrs. The reaction mixture is worked upon is the case n-butyl acetate. The yield is 58%, b.p. 111-112°.

$$CH_3CH_2CH(OH)CH_3 + CH_3COOH \xrightleftharpoons{HCl\ (g)} CH_3CH_2CH(CH_3)-COOCH_3$$

Butan-2-ol — Acetic acid — Sec. Butyl acetate

Diethyl Adipate

It is obtained by refluxing for 5-6 hr, mixture of adipic and (20 g, 0.137 mol), absolute alcohol (20 mL), dry benzene (50 mL) and conc. H_2SO_4 (4 mL) in a R.B. flask (200 mL capacity) using a Dean-stark apparatus. The mixture is poured into water (150 mL), benzene layer separated and the aqueous solution extracted with benzene (50 mL). The combined benzene extract is dried ($CaCl_2$) and distilled. Diethyl adipate is obtained in 94% yield, b.p. 134-135°/17 mm.

$$\underset{\text{Adipic acid}}{\begin{array}{l} CH_2CH_2COOH \\ | \\ CH_2CH_2COOH \end{array}} + C_2H_5OH \xrightarrow[\text{dry benzene reflux}]{H_2SO_4} \underset{\text{Diethyl adipate}}{\begin{array}{l} CH_2CH_2COOC_2H_5 \\ | \\ CH_2CH_2COOC_2H_5 \end{array}} + 2H_2O$$

Dimethyl Adipate

It is obtained by stirring a mixture of adipic acid (29. 25 g, 0.2 mol), dry methanol (10 mL), 2,2-dimethoxypropane (41.6 g, 0.4 mol) and p-toluenesulphonic acid (0.25 g) in a R.B. flask (250 mL capacity) fitted with a reflux condenser. The mixture is heated at 45° (water bath) for 4 hrs. The acetone and methanol are distilled (water bath). The residual dimethyl adipate is distilled, b.p. 130°/25 mm, yield 84%.

$$\underset{\text{Adipic acid}}{HOOC(CH_2)_4COOH} + 2CH_3OH \xrightarrow{H^+} \underset{\text{Dimethyl adipate}}{CH_3OOC(CH_2)_4COOCH_3} + 2H_2O$$

$$\underset{\text{2, 2-Dimethoxy propane}}{CH_3C(OCH_3)_2CH_3} + H_2O \longrightarrow \underset{\text{Acetone}}{CH_3-\overset{\overset{\displaystyle O}{\|}}{C}-CH_3} + 2CH_3OH$$

Ethylphenyl Acetate

It is obtained by refluxing for 4 hr., a mixture of benzyl cyanide (14.8 mL, 0.13 mol), rectified spirit (30 mL) and conc. H_2SO_4 (13.6 mL) is a R B flask (100 mL capacity), fitted with a reflux condenser. The mixture is poured into water (75 mL) and extracted with ether (35 mL). The ether extract is washed with sodium bicarbonate solution (4 × 10 mL) and then with water. The remaining solution in dried (anhyd. $MgSO_4$) and distilled to give ethyl phenyl acetate, b.p. 227°, yield 75%.

Ethyl p-aminobenzoate

Also known as **benzocaine**, is prepared by stirring a mixture of p-amiobenzoic and (12 g, 0.088 mol) and absolute ethanol (100 mL) is a R B flask until a clear solution is obtained. To the stirred solution is added dropwise conc. H_2SO_4 (4 mL). The formed precipitate dissolves on subsequent refluxing. The mixture is refluxed at about 105° for 1.5 hr while stirring. To the cooled reaction mixture is added water (50 mL) followed by dropwise addition of sodium carbonate solution (10%) to neutralise the acid. More sodium carbonate solution is added until pH is about 8. The separated benzocaine is filtesd, washed with water and crystallised from dilute methanol M.P. 92°. Yield 60%.

$$H_2N-C_6H_4-COOH + C_2H_5OH \xrightleftharpoons{H^+} H_2N-C_6H_4-COOC_2H_5$$

p-Aminobenzoic acid — Ethyl p-aminobenzoate (Benzocane)

Alternatively, Ethyl p-aminobenzoate can also be obtained by passing dry hydrogen chloride gas into absolute alcohol (80 mL) till the alcohol is completely saturated. To this solution is added p-aminobenzoic acid (12 g, 0.088 mol) and the mixture refluxed for 2 hr on a water bath (anhydrous conditions). The hot solution is poured into water (150 mL) and sodium carbonate is added until the solution is neutral to litmus. The precipitated ethyl p-aminobenzoate is filtered and crystallised form ethanol, M.P. 91°, yield 69%.

Isopentyl Acetate

Also known as BANNARA OIL due to its familiar odor of bannana, is preprepard by refluxing isopentyl alcohol (12.5 mL) glacial acetic acid and conc. H_2SO_4 (2.5 mL added with caution) for 1 hr. Water (25 mL) is added to the cooled solution and extracted with ether (25 mL). The ether extract is washed with sodium bicarbonate solution (2 × 5 mL), water, dried (Na_2SO_4) and distilled. Isopentyl acetate b.p. 142° is obtained in about 80% yield.

$$CH_3-CH(CH_3)-CH_2CH_2OH + CH_3COOH \xrightleftharpoons{H^+} CH_3-CH(CH_3)-CH_2CH_2O-C(=O)-CH_3$$

Isopentyl alcohol — Acetic acid — Isopentyl acetate

In addition to the above esters, the following esters are also of importance and prepared in a similar way.

Ethylbutyrate (odour of pine apple) $CH_3CH_2CH_2CH_2C(=O)-OCH_2CH_3$

Isobutylpropionate (odour of rum) $CH_3CH_2-C(=O)-OCH_2CH(CH_3)_2$

Octylacetate (odour of oranges) $CH_3C(=O)O-CH_2(CH_2)_6CH_3$

Methylanthrailate (odour of grapes)

$$\text{m-}H_2N{-}C_6H_4{-}\overset{O}{\overset{\|}{C}}{-}OCH_3$$

Benzylacetate (odour of peach)

$$C_6H_5CH_2-O-\overset{O}{\overset{\|}{C}}-CH_3$$

n-Propylacetate (odour of pears)

$$CH_3CH_2CH_2O\overset{O}{\overset{\|}{C}}-CH_3$$

Methylbuyrate (odour of apples)

$$CH_3CH_2CH_2CH_2\overset{O}{\overset{\|}{C}}-OCH_3$$

Ethyl phenylacetate (odour of honey)

$$C_6H_5CH_2-\overset{O}{\overset{\|}{C}}-OCH_2CH_3$$

12.14 ETHYL RESORCINOL

It is obtained by **Clemmensen Reduction**[1] of resacetophemone with zinc amalgam.

$$\text{Resacetophenone (HO, OH, }COCH_3\text{)} \xrightarrow[HCl]{Zn-Hg} \text{Ethyl resorcinol (HO, OH, }CH_2CH_3\text{)}$$

Resacetophenone Ethyl resorcinol

Procedure

A mixture of amalgamated zinc (obtained form 20 g. zinc, see note 2) and resacetophenon (5 g) is refluxed is a R.B. flask for about 3 hr and conc. HCl (2.5 mL) is added to it every hour. The solution is cooled, saturated with solid sodium chloride (table salt) and extracted with ether (5 × 20 mL). The combined ether extract is dried (Na_2SO_4) and ether distilled to give ethyl resorcinol, m.p. 97°, yield 4 g (88%)

Notes

1. W.T. Borden and T. Ravindranathan, J. Org. Chem., 1971, **36**, 4125; T. Nabayaski, J. Am. Chem. Soc., 1960, **82**, 3900, 3906, 3909)

2. Zinc amalgam is prepared as follows:

   Zinc turnings (20 g) is washed with dil. HCl (1.1). To the washed zinc is added mercuric chloride (1.5 g), conc. HCl (1 mL) and water (20 mL). The mixture is vigorously stirred for 10 min. aqueous solution decanted and amalgamated zinc washed of with distilled water. It is covered with dil HCl (30 mL 1:1) and used for the reduction

3. Resacetophenone is obtained by **Nencki reaction.**

   Freshly fused and powdered $ZnCl_2$ (16.5 g) is dissolved is glacial acetic acid (16 mL) by heating (sand bath) in a beaker. Dry resorcinol (11 g) is added with stirring to the mixture at 140°. The solution is heated at 150° for 20 min. Dil. HCl (1:1, 5 mL) is added to the mixture and the solution cooled (5°). The separated resacetophenone is filtered, washed with dil. HCl (1: 3) and crystallised from hot water, m.p. 142-143°, yield 14 g (93%)

   Resacetophenore can also be obtained by **Houben-Hoesch reaction**

4. Clemmensen reduction of benzil gives dibenzyl, m.p. 52° in 84% yield

$$\underset{\text{Benzil}}{C_6H_5-\overset{\overset{\displaystyle O}{\|}}{C}-\overset{\overset{\displaystyle O}{\|}}{C}-C_6H_5} \xrightarrow[\text{HCl}]{\text{Zn/Hg}} \underset{\text{Dibenzyl}}{C_6H_5-CH_2CH_2C_6H_5}$$

12.15 HYDROXYHYDROQUINONE TRIACETATE

It is obtained by **Thiels Acetylation** of hydroquinone by reaction with acetic anhydride in presence of conc. H_2SO_4

p-Benzoquinone + $(CH_3CO)_2O$ (Acetic anhydride) $\xrightarrow[\text{1, 4-addn.}]{H^+}$ [intermediate (C=O, OH, H, $OCOCH_3$) $\longrightarrow$ dihydroxyphenyl acetate (OH, OH, $OCOCH_3$)] $\xrightarrow{(CH_3CO)_2O}$ Hydroxyhydroquinonetriacetate (three $OCOCH_3$)

Materials

p-Benzoquinone	4 g
Acetic anhydride	12 mL
Conc. H_2SO_4	0.5 mL

Procedure

To a stirred mixture of p-benzoquinone (4 g, 0.037 mol) and acetic anhydride (12 mL) is added dropwise conc. H_2SO_4 (0.5 mL) in a RB flask (100 mL capacity). The temperature of the reaction mixture is maintained between 40-50° during the addition. The stirring is continued for 5 min. and the reaction mixture poured on to crushed ice (50-75 g) The separated hydroxyhydroquinone triacetate (1,2,4-triacetoxybenzere) is filtered, washed with water and crystallised form alcohol, m.p. 96-97°, yield 8.5 g (91.1%)

Note

1. The required p-benzoquinone is obtained by the oxidation of hydroquinone with Pot. bromate (V.K. Ahluwalla, R. Aggarwal, Comprehensive Practical Organic Chemistry: Preparation and Quantitative Analysis, Universites Press, 2000, Page 169).

12.16 HYDROQUINONE DIACETATE

It is obtained by the reaction of hydroquinone with acetic anhydride in presences of conc. H_2SO_4

$$HO-C_6H_4-OH + (CH_3CO)_2O \xrightarrow{Conc.H_2SO_4} H_3COCO-C_6H_4-OCOCH_3$$

Hydroquinone; Acetic anhydrode; Hydroquinone diacetate

Materials

Hydroquinone	2.8 g
Acetic anhydride	5 mL
Conc. H_2SO_4	1-2 drops

Procedure

Conc. H_2SO_4 (1-2 drops) is added to a mixture of hydroquinone (2.8 g, 0.025 mol) and acetic anhydride (5 mL) is a conical flask. The reaction is exothermic, mixture gets heated up and a clear solution results. The reaction mixt. is kept for 5 min. and poured on to crushed ice (40-45 g). The separated hydroquinone diacetate is filtered, washed with water and crystallised form dilute alcohol (1 : 1), m.p. 122°, yield 4.6 g (95%).

Notes

1. The reaction can also be performed by irradiation with microwave for 30 sec.
2. Hydroquinone diacetate can also be prepared by heating p-benzoquinone and acetic anhydride in presence of zinc dust and anhydrous sodium acetate.

12.17 METHYL CYCLOPENTANE CARBOXYLATE

It is obtained by **Favorskii Rearrangement**[1] involving the reaction of 2-chlorocyclohexanone with sodium methoxide.

H $\overset{\ominus}{O}CH_3$

=O ⟶ $CH_3OH + Cl^-$ + =O $\xrightarrow{\overset{\ominus}{O}CH_3}$

Cl

2-Chlorocyclohexanone

⟶ O^- OCH_3 ⟶ $\ominus$ O C OCH_3 $\xrightarrow{CH_3OH}$ $COOCH_3$

Methyl cyclopentane carboxylate

Materials

2-Chlorocyclohexanone	8.25 g
Sodium methoxide	3.62 g
Dry ether	30 mL

Procedure

A solution 2-chlorocyclohexanone (8.25 g, 0.062 mol) in dry ether (10 mL) is added dropwise to a stirred suspension of sodium methoxide (3.62 g, 0.067 mol) in dry ether (20 mL) during 40-45 min. The reaction is carred out in a three necked RB flask fitted with a reflux condenser and a dropping funnel (all filled with $CaCl_2$ guard tubes). The stirring is continued and the mixture refluxed for 2 hr (water bath). It is cooled, water (20 mL) added and the ether layer separated. The aqueous layer is saturated with sodium chloride and extracted with ether (2 × 10 mL). The combined ether solution is washed with dil. HCl (1 : 1, 10 mL), sodium bicarbonate (15%, 10 mL) and sat'd NaCl solution (10 mL). It is dried ($MgSO_4$) and ether evaporated to give methyl cyclopentane carboxylate 4.5 g (57%), b.p. 70-73°/48 mm.

Notes

1. A.E. Favorski, J. Prakt. Chem., 1913, **88(2)**, 658, O. Wallach, Ann., 1918, 414, 296.

12.18 α-NAPHTHYL ACETIC ACID

It is obtained by the treatment of α-napnthoyl chloride with diazomethane followed treatment of the formed α-naphthyl diazoketone with silver oxide in presence of sodium thiosulphate. The reaction is known as **Arndt-Eistert synthesis**[1]

$$\alpha\text{-Naphthoyl chloride (COCl)} \xrightarrow{CH_2N_2} \text{Didzoketone (COCHN}_2\text{)} \xrightarrow[Na_2S_2O_3]{Ag_2O} \alpha\text{-Naphthyl acetic acid (CH}_2\text{COOH)}$$

α-Naphthoyl chloride   Didzoketone   α-Naphthyl acetic acid

The mechanism involved is given below:

$$R-\overset{O}{\overset{\|}{C}}-Cl + \overset{\ominus}{C}H_2-\overset{\oplus}{N}\equiv N \longrightarrow R-\overset{O}{\overset{\|}{C}}-CH_2-\overset{\oplus}{N}\equiv N + Cl^-$$

$$R-\overset{O}{\overset{\|}{C}}-\underset{H}{CH}-\overset{\oplus}{N}\equiv N + \overset{\ominus}{C}H_2-\overset{\oplus}{N}\equiv N \longrightarrow R-\overset{O}{\overset{\|}{C}}-\overset{\ominus}{C}H-\overset{\oplus}{N}\equiv N + CH_3\overset{\oplus}{N}\equiv N$$

$$\downarrow -N_2$$

$$\underset{\text{ketene}}{R-CH=C=O} \longleftarrow \underset{\text{Carbene}}{\left[R-\overset{O}{\overset{\|}{C}}-\ddot{C}H\right]}$$

$$\underset{\text{Ketene}}{R-CH=C=O} + H_2\ddot{O} \longrightarrow R-CH_2-\overset{O}{\overset{\|}{C}}-OH$$

Procedure

α-Naphthoyl chloride (5 g) in dry ether (15 mL) is added dropwise to an ethereal solution of diazomethane at 10°. The reaction mixture is kept for 6 hr. at 20-25° and ether removed at 30° (in vacuo) to give α-naphthyl drazoketone which crystallised from benzene, yield 5 g (98%), m.p. 54-55°.

A solution of the above diazoketore (3 g) in dry dioxane (25 mL) is added to a suspension of silver oxide (2 g), sodium thiosulphate (3 g) in water (100 mL) at 50-60° with stirring. The mixture is stirred for 1 hr, solution cooled and acidified (dil HNO_3). The separated α-naphthylacetic acid is filtered and crystallised form water, m.p. 130°, yield 50%.

Notes

1. F. Arndt and B. Eistert, Ber., 1935, 68, 200, A.B. Smith, Chem. Commun., 1974, 695.
2. The required α-Naphthoyl chloride (b.p. 168°/10 mm) is obtained form α-naphthoic acid by heating with phosphorus pentachloride.

3. **Diazomethane** required is generated by the PTC method, involving reaction of hydrazine with chloroform in alkaline solution in presence of a PTC (tetrabutylammonium hydroxide) (D. T. Sepp, K. V. Schever and W. B. Weber, Tetrahedron lett., 1974, 2983, H. Staudinger and O. Kufer, Ber., 1912, **45**, 501)

$$NH_2NH_2 + CHCl_3 + NaOH \xrightarrow[\text{etter or } CH_2Cl_2]{\text{PTC}} CH_2N_2 \text{ in ether or } CH_2Cl_2$$

Use of crown ether in place of the above PTC gives much bether yield of diazomethane. The function of PTC is the generation of carbene in situ.

12.19 2-PHENYLQUINOLINE-4-CARBOXYLIC ACID (ATOPHAN)

It is obtained by heating benzaldehyde, pyruvic acid and aniline is absolute alcohol. The reaction is known as **Doebner Reaction**[1]

$$\underset{\text{Benzaldehyde}}{C_6H_5CHO} + \underset{\text{Pyruvicaid}}{CH_3-\overset{O}{\overset{\|}{C}}-COOH} \longrightarrow C_6H_5-CH{=}CH-\underset{\underset{O}{\|}}{C}-COOH \xrightarrow{H_2\ddot{N}C_6H_5}$$

O, COOH, C=CH, CH — C_6H_5, ⊕ N, H, H $\underset{+H^+}{\overset{-H^+}{\rightleftharpoons}}$ O, COOH, C, N̈, C_6H_5, H $\longrightarrow$ Ō, COOH, +N, C_6H_5, H $\longrightarrow$

$\underset{+H^+}{\overset{-H^+}{\rightleftharpoons}}$ HO, COOH, N, C_6H_5, H $\xrightarrow{-H_2O}$ COOH, N, C_6H_5, H $\xrightarrow{-H_2O}$ COOH, N, C_6H_5

2-Phenylquinoline-4-carboxylic acid

Materials

Aniline	5.6 mL
Benzaldehde	6 mL
Pyruvic acid	5.7 mL
Absolute alcohol	100 mL

Procedure

To a boiling solution of pure benzaldehyle (6 mL, 0.059 mol), freshly distilled pyruvic acid (5.7 mL, 0.06 mol) and absolute alcohol (50 mL) (taken in a R.B. flask, fitted with a reflux condenser and $CaCl_2$ guard tube) is added slowly with shaking, a solution of pure aniline (5.6 mL, 0.061 mol) in absolute alcohol (50 mL). The mixture is refluxed for 3 hr and allowed to stand overnight. The separated 2-phenylquinoline-4-carboxylic acid is filtered, washed with cold water and crystallised form alcohol, m.p 210°, yield 7.5 g (51%)

Notes

1. O. Doebner, Ann. 1887, **242**, 264, Ber, 1887, **20**, 277; Ber, 1894, **27**, 352, 2020
2. Pure **pyruvic acid** is obtained as follows. An intimate mixture of powdered tartaric acid (25 g, 0.16 mol) and freshly fused potassium hydrogen sulphate (37.5 g, 0.26 mol) is heated in a RB flask (250 mL capacity) with a distillation arrangement. The reaction mixture is heated at 210-220° (oil-bath) till no more liquid distils. The distillate is redistillled under reduced pressure. The yield is 7.5 g (51%), b.p. 75-80°/25 mm.

12.20 REDUCTION OF ALDEHYDES TO ALCOHOL

Aldehydes are known to be reduced to primary alcohols by Na-EtOH, 2-4 % PtO_2/H_2; 5% Pd/C, Raney Ni, $LiAlH_4$, $NaBH_4$, HCO_2H/EtMgBr, (EtO)Si HMe and enzymatically (For details sec V.K. Ahluwalia and R.K. Parashar, Organic Reaction Mechanism, Narosa Publishers, 2007, Page 257-258).

The aldehydes, can be reduced to primary alcohols by **hydrogen-Transfer reactions**. The procedure involves healing aldehydes with neat iso PrOH, 220-250°, for 24-30 hr. The reaction is best conducted in a sealed tube (J. Haggin, Chem. Enq. News, 1994, **72(2)**, 22, J.K.S. Wan and T.A. Koch, Res. Chem. Intermed, 1994, 20, 29; M. Mehdizadch, Res. Chem. Intermed, 1994, **20**, 79). Following aldehydes have been reduced.

CHO → ($\xrightarrow[\text{iSO PrOH}]{225°/24\ \text{hr}}$) CH_2OH

Benzaldehyde — Benzyl alcohol (92%)

CHO → ($\xrightarrow[\text{iSO-PrOH}]{227°/29\ \text{hr}}$) OH

Trans-cinnamylaldehyde — Cinmanyl alcohol (83%)

CHO, MeO → ($\xrightarrow[\text{iSO-PrOH}]{225/27\ \text{hr}}$) OH, MeO

p-Methoxy benzaldehyde — p-Methoxybenzyl alcohol (81%)

12.21 β-RESACETOPHENONE (2, 4-DIHYDROXYACETOPHONONE)

It obtained by **Nencki Reaction** involving reaction of resorcinol with acetic acid in presence of anhyd. $ZnCl_2$

$$\text{Resorcinol (HO, OH)} \xrightarrow[\text{anhyd. } ZnCl_2, \Delta, 140°]{CH_3COOH} \beta\text{-Resacetophenone (HO, OH, } COCH_3\text{)}$$

Resorcinol β-Resacetophenone

Materials

Resorcinol	5.5 g
Anhyd. $ZnCl_2$	8.25 g
Glacial Acetic Acid	8 mL

Procedure

Dry resorcinol (5.5 g, 0.05 mol) is added to a solution (obtained by heating on a sand bath in a beaker) of freshly fused and powdered $ZnCl_2$ (8.25 g, 0.13 mol). The mixture is stirred with a glass rod at 140°. The solution is heated unit it just begins to boil and kept for 20 min. at 150°. Dil. HCl (1 : 1, 30 mL) is added to the mixture and solution cooled (5°). The separated β-reacetophenone is filtered, washed with dil. HCl and crystallised from hot water, m.p. 142-144°, yield 7 g (93%).

12.22 TOLAN (DIPHENYLETHYNE)

It is obtained by the oxidation of benzildihydrazone with yellow mercury oxide.

$$C_6H_5-\underset{\underset{N-NH_2}{\|}}{C}-\overset{\overset{N-NH_2}{\|}}{C}-C_6H_5 \xrightarrow[2HgO]{[O]} C_6H_5-C\equiv C-C_6H_5 + 2N_2 + 2H_2O + 2Hg$$

Benzil dihydrazone

Materials

Benzil dihydrazone	5 g
Benzene	30 mL
Yellow mercuric oxide	12 g

Procedure

Yellow mercuric acid (obtained by adding sodium hydroxide solution to a solution of mercuric chloride) is added to a suspension of benzil dihydrazone (5 g, 0.02 mol). The reaction mixture is warmed (water-bath). Evolution of nitrogen takes place and mixture turns grey. The remaining yellow mercuric oxide (7 g) is added (is small lots) so as to keep the reaction mixture gently refluxed. The reaction mixture is gently refluxed for 1 hr, left overnight and filtered. The residual mercuric oxide is washed with benzene (5 mL). The combined benzene solution is dried (Na_2SO_4) and distilled. The fraction between 95-105° is collected. It solidified on cooling and crystallised from alcohol m.p. 60-61°. yield 24 g (67.4%).

Notes

1. The required benzil dihydrazone, m.p. 150-151° is obtained by adding hydrazine hydrate (85%, 6.1 mL) to a solution of benzil (8.4 g, 0.04 mol) in n-propyl alcohol (25 mL). The mixture is refluxed for a total of 50 hrs. The separated benzil dihydrazone is filtered and crystallised form alcohol yield 93.7%

12.23 TRIMETHYL ACETIC ACID (2,2-DIMETHYL PROPIONIC ACID)

It is obtained by the haloform reaction (oxidation of pinacolone with sodium hypobromite)

$$\underset{\text{Pinacolone}}{(CH_3)_2C - \underset{\underset{O}{\|}}{C} - CH_3} \xrightarrow[(2)\ H_3^+O]{(1)\ NaOBr} \underset{\text{Trimiethyl aceticacid}}{(CH_3)_3CCOOH + CHBr_3}$$

$$H\bar{O} + Br_2 \rightleftharpoons :\ddot{\bar{O}}\,Br + :\bar{B}r + H_2O$$

$$RCOCH_3 + :\ddot{\bar{O}}\,Br \rightleftharpoons [RCO\ddot{C}H_2]^- + HOBr$$

$$RCO\overset{\ominus}{\ddot{C}}H_2 + Br - OH \longrightarrow RCOCH_2Br + :\overset{\ominus}{O}H$$

$$RCOCH_2Br \longrightarrow RCOCBr_3 \xrightarrow{:\bar{O}H} RCO\bar{O} + CHBr_3$$

$R = (CH_3)_2C -$

Materials

Pinacolone	20 g
Sodium hydroxide	64 g
Bromine	31.3 mL

Procedure

Bromine (31.3 mL, 0.6 mol) is added dropwith to a stirred and cooled (0°) solution of sodium hydroxide (64 g, 1.6 mol) in water (280 mL). The mixture is well stirred and the temperature is not allowed to rise above 10° (15-20 min). The mixture is cooled to 0° (ice-bath) and pinacolone (20 g, 0.2 mol) is added (temp. is kept below 10°). The mixture is stirred for 3 hr at room temperature. The reaction mixture is steam distilled to remove carbon tetrachloride and bromoform. It is cooled and conc. H_2SO_4 (80 mL) added. The reaction mixture is heated again in the distillation assembly. Trimiethyl acetic acid passes over with about 100 mL water. The upper lager of trimethyl acetic acid is separated (separatory funnel), dried (Na_2SO_4) and distilled. Yield 13.2 g (55%), b.p. 75-80°/20 mm, m.p. 34-35°

Notes

1. Sodium hypochlorite can also be used for the oxidation of mesityl oxide to 3,3-dimethylacrylic acid
2. See also preparation of 2-Naphthoic acid (Section 2.33)

Index

C

D

E

T

U

V

W